Philosophy of Physics

PRINCETON FOUNDATIONS OF CONTEMPORARY PHILOSOPHY

Scott Soames, *Series Editor*

Philosophical Logic by JOHN P. BURGESS
Philosophy of Language by SCOTT SOAMES
Philosophy of Law by ANDREI MARMOR
Truth by ALEXIS G. BURGESS & JOHN P. BURGESS

PHILOSOPHY OF PHYSICS
Space and Time

Tim Maudlin

PRINCETON UNIVERSITY PRESS
PRINCETON AND OXFORD

Published by Princeton University Press,
41 William Street, Princeton, New Jersey 08540
In the United Kingdom: Princeton University Press,
6 Oxford Street, Woodstock, Oxfordshire OX20 1TW
press.princeton.edu

Fifth printing, and first paperback printing, 2015

Paperback ISBN 978-0-691-16571-4

The Library of Congress has cataloged the cloth edition as follows

Maudlin, Tim.
    Philosophy of physics : space and time / Tim Maudlin.
        p. cm. — (Princeton foundations of contemporary philosophy)
    Includes bibliographical references and index.
    ISBN 978-0-691-14309-5 (alk. paper)
    1. Space and time. I. Title.
    BD632.M38 2012
    530.11—dc23   2012005618

British Library Cataloging-in-Publication Data is available

This book has been composed in Minion Pro & Archer

Printed on acid-free paper. ∞

Printed in the United States of America
10 9 8 7 6 5

To the memory of Robert Weingard,
seeker of the ways of the universe

Contents

Contents

Acknowledgments

THE ROOTS OF THIS VOLUME stretch back in time to my graduate career, when Clark Glymour graciously agreed to teach a year-long seminar on Relativity at the request of the graduate students in the History and Philosophy of Science at Pitt. Without that patient and thorough presentation of modern mathematical methods, I would not have been able to start thinking about the theory in a fundamentally geometrical way. John Norton's seminars opened up the historical background going back to Newton, and Peter Machamer guided us through Galileo. John Earman and Norton had recently articulated the hole argument, and I cut my teeth puzzling over that conundrum. In sum, this book is the result of a quarter century of reflection on the teeming profusion of ideas that was the lifeblood of that remarkable program.

When I came to Rutgers in 1986, I was immeasurably fortunate to have Robert Weingard as a friend and colleague. His curiosity and intellectual honesty always made discussions a delight, and I profited from his deep knowledge of physics. This volume is dedicated with profound gratitude and affection to him.

I owe a different sort of debt to the many students, both graduate and undergraduate, whom I have had the privilege to teach over these years. The presentation of space-time theory found here has slowly evolved over many classes. At first I followed standard presentations, making extensive use of coordinates and coordinate transformations. Bit by bit, class after class, reference to coordinates dropped away, leaving the fundamental geometry open to direct inspection. The presentation of Relativity, in particular, is somewhat unorthodox but (knock wood) conceptually clear. I hope at least to save the reader from a few of the confusions that I had to struggle to overcome.

The manuscript has benefitted from feedback and comments from many quarters. I am particularly grateful to two anonymous referees: I hope they find the final version improved from the one they perused. Sean Carroll was correctly insistent that I get the details right on General Relativity, and chapter 6 is much improved due to his advice. Adam Elga provided helpful comments, as did the philosophy of science reading group at NYU. The indefatigable Bert Sweet worked through the details, and the calculations, with his trademark care and attention.

I am also grateful to Scott Soames, Rob Tempio, and Princeton University Press for allowing me to expand what was meant to be a single volume on the philosophy of physics into two. It is painful to consider the compromises and inadequacies that a single volume would have required.

At a more practical level, the time needed to complete the manuscript was afforded by a sabbatical semester at Rutgers. *Merci.*

Finally, as always, Vishnya Maudlin has done much more than put up with the obsession that accompanies writing a book. She has always been there, in class and at home, happy to discuss and criticize and clarify. I cannot imagine taking on a task like this as a solitary pursuit. Without her, it would not have come to be.

Introduction: The Aim and Structure of These Volumes

PHILOSOPHY OF PHYSICS concerns the whole of physical reality, considered in a usefully generic way. For example, the physical world appears to have spatial and temporal aspects, so the existence and nature of space and time (or space-time) is a central topic. Matter, the sort of stuff of which tables and chairs and planets are composed, is a similarly central topic. By "a usefully generic way," I mean this: the most general question we can ask about matter is what sort of thing it is. For example, we might hold that matter is made of point-like particles, or of fields, or of one-dimensional strings, or of some combination of these, or of something else altogether. Given any one of these general accounts, there are further, more specific questions: how many sorts of fields there are, what the masses of the particles are, and so on. We will be concerned with the most general questions, rather than the more specific ones.

Philosophy of physics, as a discipline, is continuous with physics proper. The sorts of questions we will ask are among the questions physicists ask, and among the questions physical theories historically have tried to answer. But an astonishing amount of physics can proceed without answers to these questions. For example, the science of thermodynamics, as its name suggests, initially aimed at providing a precise mathematical account of how heat spreads through an object and from one object to another. But we can discover quite detailed equations governing heat flow and still not have an account of what heat *is*. Is it a sort of fluid (as caloric theory holds) that literally flows out of object and into another, or a sort of motion (as kinetic theory holds) that is communicated by interaction from one body to the other? If all you care about is how long it will take a 20-pound iron rod at 200° F to cool to 100° F when it is immersed in a large vat of water at 50° F, the equations of heat flow can provide the answer. But you will be none the wiser, having calculated the answer, about the fundamental nature of heat. An ironworker may not give a fig about the nature of heat, and the philosopher of physics may care equally little about the exact time it takes for the iron to cool down. A practicing physicist will typically care about both but may focus more on one or the other at different times. It is characteristic of a contemporary physics education that much more time is spent learning how to solve the equation and get a practical answer for the ironworker than in discussing the more "philosophical" questions about the nature of heat, or the nature of space and time, or the nature of matter. Physics students who are fascinated by these more foundational questions can find themselves frustrated by physics classes that refuse to address them. This volume is dedicated to them as much as it is to philosophers with an interest in physical reality.

The philosophy of physics has here been divided into three parts, spread over two volumes. Each of these volumes can be read independently of the other. But particular themes—most importantly, the need for a completely physical account of "measurement" procedures—are addressed in both volumes, so reading them in order will repay the effort. The first volume addresses the nature of space and time. It contains a brief history of debates about space and time from classical physics (Newton) through General Relativity. In physics, space and time (or later, space-time) serve as the stage on which the history of the physical universe plays out. But space and time themselves are elusive entities. The physical world presents itself to us as a collection of things and events in space that coexist or succeed each other in time. But space and time themselves do not appear to our senses: they have no color or flavor or sound or smell or tangible shape. What space and time seem rather to have is a geometrical structure. We will examine various theories about exactly what that structure *is*, and about what *has* that structure. The Theory of Relativity is presented, first and foremost, as a theory of the geometry of space-time. Special Relativity is explained in enough detail to solve specific problems about the behavior of clocks and rigid objects in a relativistic world. General Relativity is presented less rigorously. My aim has been to make the conceptual foundations of these theories absolutely clear, with particular attention to how the use of coordinates in physics relates to the underlying geometrical structure.

Volume 2 takes up the theory of matter. The first part of this volume presents the contemporary theory of matter: quantum theory. Unlike Relativity, there is no agreement among physicists about how to understand quantum theory. Indeed, the very phrase "quantum theory" is a misnomer: there is no such theory. Rather there is a mathematical formalism and some (quite effective) rules of thumb about how to use the formalism to make certain sorts of predictions. Here the difference between the ironworker and the philosopher of physics becomes acute. The ironworker (or the physicist in ironworker mode) doesn't particularly care about the nature of the physical reality: it is enough to calculate how various experiments should come out. The philosopher of physics cares about the underlying reality and attends to the predictions only insofar as they can serve as evidence for which account of the underlying reality is correct. In this part, we will consider some competing accounts of the nature of matter. These theories share much of the mathematics of quantum theory in common but nonetheless differ radically in their accounts of what exists.

If volume 1 covers space-time and the first part of volume 2 covers the material contents of space-time, it might seem that there is nothing more to discuss. Haven't we licked the platter clean? In a sense we have: all there is to the physical world, at a fundamental level, is accounted for by the theory of space-time and the theory of matter. Nonetheless, there are physical phenomena that are more perspicuously understood and explained by using a different set of concepts than those peculiar to space-time theory and quantum theory. A signal example of this is thermodynamics. Even though the phenomena addressed by thermodynamics are, at base, nothing more than the motions of matter in space-time, still a certain sort of insight, understanding, or explanation requires analyzing them with the conceptual tools of statistical mechanics. These same tools shed light on the appearance of probabilities in physics, on the explanation of statistical patterns of behavior, and on the apparent irreversibility or time-asymmetry of many phenomena. Our investigation of the relationship between thermodynamics, entropy, statistical mechanics, and irreversibility provides an example of how new insights into physical phenomena can be found even when the fundamental ontology and its laws are already known.

These books present an opinionated survey. There is far too much material and too little space to do justice to all the physical theories and philosophical positions that have been offered on these topics, and I make no pretense to try. Rather, I have considered a circumscribed set of alternative approaches that strike me as both clear and instructive. And I unabashedly advocate those that I think are the most promising and well founded. This is not a dispassionate overview of the field. But I hope that my selection of proposals illustrates what it is for a physical theory to be clear and comprehensible. Unfortunately, physics has become infected with very low standards of clarity and precision on foundational questions, and physicists have become accustomed (and even encouraged) to just "shut up and calculate," to consciously refrain from asking for a clear understanding of the ontological import of their theories. This attitude has prevailed for so long that we can easily lose sight of what a clear and precise account of physical reality even looks like. So whether or not you are attracted by the physical theories I will discuss (and many physicists will find them distasteful), I hope you come away appreciating at least their *intelligibility*. Whether these theories are correct or incorrect, insightful or wrongheaded, we know what they are claiming about the physical world. Physicists and philosophers must demand such clarity if we are to ever understand the universe we inhabit.

Philosophy of Physics

Classical Accounts of Space and Time

THE BIRTH OF PHYSICS

The intellectual tradition that produced modern theoretical physics begins in ancient Greece. Babylonian and Egyptian astronomers compiled voluminous and accurate data about the visible positions of the sun and the planets and created mathematical models that could predict phenomena such as eclipses. But the Greek nature-philosophers introduced a novel strand of speculative theorizing into this observational enterprise. Thales, Anaxagoras, and Democritus, for example, each offered conjectures about the ultimate structure of matter: that material objects are all derived from water; are mixtures of earth, air, fire, and water; or are composed of an infinite variety of differently shaped atoms. The observable behavior of familiar objects was then explained in terms of this material constitution. According to Democritus, sweet things are composed of smooth, rounded atoms; sour things are composed of angular atoms; and so on. The notion that the perceptible properties and behaviors of large objects should be accounted for by the structure and nature of their imperceptible parts underlies physics to this day.

Aristotle provided this speculative enterprise with its name. The term "physics" derives from the Aristotelian text *Physike Akroasis*: Lectures on Nature. In Greek, *physis* refers to the nature of a thing, and Aristotle defined the nature of an object as an internal source of motion and rest that belongs to an object primarily and properly and nonaccidentally (*Physics* 192²20–23). Thus, for Aristotle, the nature of an object is revealed by how the object moves, and stops moving, when left entirely to its own devices. Release a rock in midair, without pushing it in any direction, and (apparently) of its own accord it starts to move downward. A bubble of air in a tank of water spontaneously rises. The rock and the air can be forced to do other things, but only under the constraint of an outside agent. Their inborn predilections to motion and rest are not attributable to outside agents and so must arise from the very nature of the thing itself.

Aristotle's definition of the "nature" of an object does not readily connect physics with the project of explaining sweetness or sourness. The emphasis instead is on change in general and on locomotion in particular. Aristotle believed that different sorts of material have different natural motions, so to describe these natural motions, he needed a way to describe and categorize motion in general. He started with the most intuitive descriptions available. The element earth's natural motion is to fall—that is, to move downward. Water also strives to move downward but with less initiative than earth: a stone will sink though water, demonstrating its overpowering natural tendency to descend. Fire naturally rises, as anyone who has watched a bonfire can attest, as does air, but with less vigor.

It is all well and good to say that a stone naturally falls, or moves downward, but what exactly does "downward" mean? It is here that Aristotle leaves common opinion and begins theoretical postulation. To move downward, according to Aristotle, is to move *toward a particular place*. The natural motion of earth, on this view, is goal-directed: the stone wants to get someplace in particular, and its spontaneous motion always takes it closer to this ultimate objective. The special place that the stone strives to reach, according to Aristotle, is the center of the universe. Aristotle conceived of the whole material cosmos as forming a sphere, whose outer surface contains the fixed stars. The celestial sphere has a unique center. The "down" direction at any place in the universe is the direction toward that central point, and an unobstructed piece of earth will naturally move down in a straight line, toward the center, until it reaches that target. If it should manage to make it all the way to the center, the piece of earth will, of its own accord, stop moving.

Similarly, "up" is the direction in space that points directly away from the center. Fire and air naturally move upward in straight lines as far as they are able, with fire displacing air if they are in competition. According to Aristotle, if the sublunary sphere (the part of the universe below the orbit of the moon) were left entirely unmolested, all of the earth, air, fire, and water would naturally segregate out into four concentric spheres: pure earth at the center, surrounded successively by concentric spherical shells of water, then air, then fire. This provides a very rough approximation to the world as Aristotle believed it to be: a spherical rocky earth covered largely with oceans, surrounded by air.

The moon and sun and planets do not fit into this scheme, so Aristotle invented a fifth sort of substance, a quintessence, called *aether*. Unlike earth, air, fire, and water, aether does not naturally engage in straight-line motion toward some goal: its natural motion is uniform circular motion around the center. This motion is most perfectly realized by the sphere of fixed stars, which spins (as far as Aristotle knew) with perfect regularity, making one complete rotation in about 23 hours and 56 minutes (a *sidereal day*). The rest of the superlunary objects—the moon, and sun, and planets—are not so regular: as they are carried about by the sphere of fixed stars, they also execute their own more complicated periodic motions. Having identified uniform circular motion as the natural state for aether, Aristotle set a problem for astronomers of succeeding generations: account for the apparent motions of the sun, moon, and planets by some compound effect of different uniform circular motions. This basic constraint on astronomical theory remained in place until Kepler proposed his first two laws of planetary motion in 1609.

Unfortunately, even a comically inadequate sketch of the history of physics and astronomy is beyond our scope. But Aristotle's innovation, his focus on natural locomotion as the primary subject of physics, shapes the field to this day. Our first order of business is to understand what exactly "locomotion" is.

The term "locomotion" wears its meaning on its sleeve: it is not just any change but change of place (*locus*). And place, for Aristotle, is location in a spatial universe with a very special shape: a sphere. Because it is a sphere, Aristotle's universe contains a geometrically privileged center, and Aristotle makes reference to that center in characterizing the natural motions of different sorts of matter. "Upward," "downward," and "uniform circular motion" all are defined in terms of the center of the universe. If Aristotle's universe did not have a spherical shape, his physics could not have been formulated.

The importance of an account of space in the formulation of physics cannot be overstated. If physics is first and foremost about motion, and motion is change of place, then (it seems) there must be *places* that material objects can successively occupy. An object rests when it occupies the same place over time, like Aristotle's stone at the center of the universe. It is tempting to say that without some sort of space in which things move, physics cannot even get off the ground. Aristotle adopts the concept of space, and the correlative concept of motion, that we all intuitively employ. He realizes that his physics requires this space to have some particular structure—a target goal that falling objects are seeking—and postulates a physical geometry that provides this structure. The resulting finite spherical universe is foreign to us today but would have felt quite familiar to any ancient Greek.

In short, space is the arena of motion, and so an account of space will play a central role in any scientific theory of motion. Abandoning Aristotle's spherical universe entails abandoning his basic physical principles and rethinking the form that the laws of physics can take. This task was undertaken by Isaac Newton.

NEWTON'S FIRST LAW AND ABSOLUTE SPACE

If we were to axiomatize Aristotle's physics, there would be different axioms for different sorts of matter: "Earth, if unimpeded, will move in a straight line toward the center of the universe" and "Aether, if unimpeded, will move in uniform circular motion about the center of the universe." Newton did present his physics as a set of axioms, which he denominated Laws of Motion. A tremendous amount of theory is packed into these laws, and it is only a slight exaggeration to say that everything we need to know about Newtonian physics is implicit in his First Law of Motion:

> Law I: Every body perseveres in its state either of rest or of uniform motion in a straight line, except insofar as it is compelled to change its state by impressed forces.[1]

[1] My translation.

This single law smashes the Aristotelian universe to smithereens.

First, Newton's law governs *every* body: stones and planets alike. Newton obliterates the distinction between astronomy and terrestrial physics, postulating a single set of principles that explains the behavior of both. We have become so accustomed to thinking of physics as possessing this sort of universality that it takes an effort to appreciate how momentous this shift is. One of the crowning moments in the argumentative structure of the *Principia* occurs when Newton calculates that the force that maintains the moon in orbit about the earth is precisely the same force that causes an apple to fall from a tree. Newton postulates a commonality of physical structure where the tradition preceding him had seen fundamental diversity.

More importantly, Newton does not ascribe any particular natural motion to a body, as Aristotle did. Rather, the Law of Inertia attributes an innate tendency of every body to *maintain* its state of motion, whatever that might be. There is no place in the universe that any body is inherently "directed toward," as a stone is directed toward the center of the universe in Aristotle's account. Newton's theory does not require that space have a special central point.

The arena of motion for Newton is rather an entity he calls *absolute space*. Motion, for Newton, is change of location in this space. The role of absolute space in Newton's theory is so deep and pervasive that it seems impossible to make sense of anything he writes without accepting its existence. We will consider various properties of absolute space, leaving the most controversial claims for last.

First, Newton assumes that absolute space possesses the geometrical structure of three-dimensional Euclidean space. We will designate this structure E^3. E^3, unlike Aristotle's physical universe, is infinite in all directions and so has no geometrical center. Figures in E^3 obey the axioms of Euclidean geometry: for example, the sum of the interior angles of any triangle equals two right angles.

It will be useful in the following discussion to distinguish several different sorts of geometrical structure, which form a hierarchy. Each level corresponds to one of the three instruments used in Euclidean geometry: the pencil, the straightedge, and the compass. What sort of geometrical structure must a space have in order for each of these instruments to operate?

The most basic, fundamental level of geometrical structure in a space is called its *topology*. The topology of a space determines facts about *continuity*. For example, when we use a pencil in making a Euclidean diagram, we are supposed to draw *continuous* lines in the space: if we were to occasionally lift the pencil and then drop it down again elsewhere while drawing what should be a single line, we will not get a single, connected, continuous line. But in order for there to be a distinction between a single line in a space and a pair of disconnected lines, the points in the space must have some geometrical organization. This level of organization is the topology of the space.

Topology is sometimes called "rubber-sheet geometry," and the name is properly evocative. Suppose some figures are drawn on a rubber sheet, and then the sheet is stretched without tearing or pasting. Many of the properties of the figures will be changed under such deformations: straight lines can be bent into curved lines; nearby points can be pulled apart so they are more distant; a triangle can be deformed smoothly into a circle; and so on. But some features of the figures will remain unchanged: if two lines intersect before the deformation, they intersect afterward; if one point is in the interior of a closed figure and another outside before the deformation, they will remain so after; and so on. The deformations are not allowed to "tear" or "paste" the space, and topology provides the level of geometrical structure that defines what counts as "tearing" and "pasting." Tearing separates some continuous lines into discontinuous pieces, and pasting joins discontinuous lines into continuous wholes. If a space did not have a topology, then there could be no distinction between drawing a single continuous curve and drawing several disconnected curves, so Euclidean constructions could not even start.[2]

---

[2] In modern mathematics, the topology of a space is specified in terms of the *open set* structure of the space, and continuity is defined in terms of the open sets. I believe that this account of continuity and hence topology is not the most perspicuous way to describe the intrinsic geometrical structure of space-time, and have developed an alternative (see Maudlin [2010] for an overview). This is not the place to fight that battle.

The second instrument of Euclidean geometry is the straight-edge (N.B.: not the *ruler*; the straightedge has no marked scale). With a straightedge and a pencil, we can draw not just continuous lines but straight lines. The first two postulates of Euclidean geometry concern the use of the straightedge and hence make implicit claims about the structure of straight lines in Euclidean space. In particular, the first two postulates state:

1. It is possible to draw a straight line from any point to any point.
2. It is possible to extend any finite straight line continuously in a straight line.[3]

In order for these postulates to obtain, there must first of all be a distinction in the space between straight lines and other lines. This distinction, which is not determined by the topology, is provided by the *affine structure* of the space. In Euclidean space, the affine structure ensures that every pair of points are the endpoints of exactly one straight line and every finite straight line can be continued indefinitely in either direction. We can describe spaces that do not have this sort of affine structure: a pair of points might determine no straight line, or more than one, or there might be a limit to how far a straight line can be produced. So Euclid's first two postulates, which describe the uses to which a straightedge can be put, already restrict the affine structure of the space he is describing.

The affine structure of a space does not determine any facts about the *lengths* of lines or the *distances* between points. This requires yet another level of geometrical form, called the *metrical structure* of the space. The compass indicates metrical structure in a space: a circle is the locus of points all equidistant from a given center. Euclid's third postulate asserts that a complete, continuous, closed circle can be drawn with any given center and radius. Again, we could imagine spaces in which this does not hold.

The hierarchical form of these three levels of structure can be illustrated by three different sorts of transformation that can be carried out on figures in Euclidean space. A *topological transformation* carries continuous lines into continuous lines. An *affine*

---

[3] My translation.

transformation must further map straight lines onto straight lines. A "uniform stretching" of the space qualifies as an affine transformation even though it changes distances between points and deforms circles into ellipses. An *isometry* is a mapping from a space onto itself that preserves distances, so circles are carried into circles.[4] Figure 1 illustrates the three sorts of mapping.

Every isometry is an affine transformation and every affine transformation is a topological transformation, but not conversely.

Modern geometry introduces another level of structure, situated between the topology and the affine structure. This is the *differentiable structure*, which distinguishes smooth continuous curves from curves with corners or sharp bends. A mapping that preserves the differentiable structure is called a *diffeomorphism* and maps smooth curves into smooth curves. While a topological transformation can map a triangle onto a circle, a diffeomorphism cannot, since a circle is smooth and a triangle has corners. The topological transformation depicted in figure 1 is a diffeomorphism: notice that the three corners of the triangle are still identifiable.

In most discussions of Euclidean geometry, the lion's share of attention goes to the Fifth Postulate. This postulate concerns the existence and properties of parallel lines. The original discovery of non-Euclidean geometries arose from attempts to prove the Fifth Postulate from the other four. Eventually, it was shown that both the Fifth Postulate and its denial are consistent with the rest of Euclid's Postulates, so there can be spaces in which straightedges and compasses behave as Euclid requires, but in which geometric figures do not have the properties that Euclid demonstrates. For example, in some non-Euclidean spaces the interior angles of a triangle sum to more than two right angles, and in others they sum to less. The Fifth Postulate plays no essential role in the formulation of Newton's physics: Newtonian mechanics could obtain in a space that contains no parallel lines at all. The existence of an affine structure and a metrical structure, on the other hand,

---

[4] While an affine transformation merely must map straight lines to straight lines, an isometry must do more than map circles to circles: the size of the circles must also be unchanged. A scale transformation, which uniformly shrinks or expands all figures, is not an isometry even though it takes circles to circles.

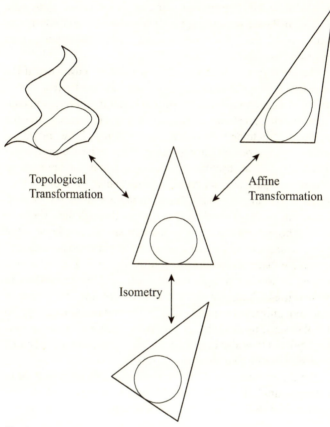

Fig. 1

is absolutely essential to make sense of Newton's Laws. But before we can make contact with those laws, we need to bring time into the picture.

Absolute Time and the Persistence of Absolute Space

Newton believed in the existence of a spatial arena with the geometrical structure of E^3. He believed that this infinite

9

three-dimensional space exists at every moment of time. And he also believed something much more subtle and controversial, namely, that *identically the same points* of space persist through time.

We are trying to understand what must be postulated if the First Law is to make sense, and the First Law asserts that a body with no forces applied to it remains at rest if it is at rest, and continues moving uniformly in a straight line if it is moving. But what is it for a body to be *resting* or to *remain at rest*? If the individual points of space persist through time, then we have a precise account: a body is at absolute rest when it occupies the same points of absolute space over a period of time. The account of absolute uniform motion in a straight line is similar but more complicated. First: if the points of absolute space persist through time, then any moving body has a trajectory in absolute space, namely the set of points in absolute space that it occupies over a given period. And if absolute space has an affine structure, then such a spatial trajectory either forms a straight line in space or it does not. Thus, to make sense of "uniform motion in a straight line," the points of space not only must persist through time and have a topology (so it makes sense to characterize a trajectory of a body as a continuous line), but they also must have an affine structure so the spatial trajectory can be characterized as straight or curved.

But these conditions alone do not define "uniform motion in a straight line," since "uniform" has not been explained. A drag racer, unlike an Indy racer, runs on a straight track, so its motion is "in a straight line." Still, the motion is not *uniform*: the drag racer accelerates, constantly moving faster and faster. This sort of motion is called *linear acceleration*. (The Stanford Linear Accelerator Center has a straight tube about two miles long down which particles are accelerated, unlike the Large Hadron Collider, which accelerates particles around a closed loop.) In order for the motion to be *uniform*, it must cover the same distance in the same time. So Newton's First Law presupposes that there is a fact about *how far* a body moves and a fact about *how much time* it takes for it to complete the motion. The first fact requires a metric on the space, so that the spatial trajectory of a body can be ascribed a length. And the second fact requires something altogether new: a metrical structure on time. Newton's First Law of Motion presupposes not only absolute space but also absolute time.

The geometrical structure of time, according to Newton (and to common sense), is simpler than that of space. Newtonian time is one-dimensional: there is a single, ordered sequence of instants that forms the totality of history. That collection of instants has a topology, which is determined by their time order. It makes no clear sense to ask whether this "time line" is straight or curved, so the notion of an affine structure does not arise. But there is a temporal metric: between any two instants a certain quantity of absolute time passes, and these quantities can be compared with each other in terms of size. If a certain amount of time passes between instant 1 and instant 2, and a certain amount of time passes between instant 2 and instant 3, there is a fact about whether these intervals are the same size or different, and a fact about what the exact ratio between the intervals is.

With all of this structure in place, we can define "uniform motion": a uniform motion is a motion that covers the same amount of space in the same time. A uniform motion need not be straight: uniform circular motion, for example, can keep a constant speed even as it continuously changes direction. So Newton needed all of these characteristics in order for the First Law to make a precise statement: when no force is put on an object, if it is at absolute rest it will remain at absolute rest, and if it is moving it will continue moving in a straight spatial trajectory, covering equal distances in equal times.

There is one last feature that Newton ascribes to absolute time: it, unlike space, has a *direction*. Newton does not make any explicit remarks about this, and it is not immediately relevant to understanding his laws of motion. But it is a perfectly natural thing to say that time passes from the past to the future, and it is worthwhile remarking here because we will return later to questions that surround the direction of time.

Indeed, Newton does not explicitly discuss the geometrical structure of space or time at all. He always *uses* E^3 as his account of space, and he always presumes in his proofs that there is a definite metric for the passage of time. It would not have occurred to him that there could be any alternative. What we have seen in the foregoing analysis is which features of his account are required in order to state the First Law: space must have a topology, an affine structure, and a metric; time must be one-dimensional with a topology and a metric; and, most importantly, the individual parts of space must persist through time. Given all this, there is a fact about whether a body remains in the same region of space through time, a fact about the spatial trajectory of a moving object, and a fact about how quickly a moving body covers different parts of that trajectory. Without this much structure, it is unclear how to make any sense of Newton's First Law. But if we were to deny that space is E^3, ascribing it instead some other affine structure and metric, the law would still make perfect sense.

The Metaphysics of Absolute Space and Time

While Newton does not make explicit note of the geometrical structure he ascribes to absolute space and time, he does provide a very clear discussion both of their metaphysical status and of the reasons he thinks we must accept their existence. The basic issue is obvious. Take, for example, an object at absolute rest at some time that is not subject to any external forces. According to the First Law, the object will remain at absolute rest—that is, it will remain located at the same place in absolute space. But Newton is perfectly aware that these persisting parts of absolute space cannot be perceived by the senses. No observation can reveal whether a body remains in the same region of absolute space or constantly moves from one part to another. It would seem, then, that even if Newton's First Law is true, and even if we could ascertain that there are no forces on an object, no observation could verify the law. And, more seriously, if we cannot perceive absolute space and *a fortiori* cannot perceive absolute motion, it is not obvious how a theory that treats of such absolute motion could make any predictions about observable fact at all.

What we can observe, Newton asserts, are the *relative* positions of bodies with respect to each other. Similarly, we cannot directly observe the passage of absolute time, but we can observe changes in the *relative* positions of bodies. In the Scholium that follows his definitions of novel terms, Newton carefully distinguishes observable quantities from the absolute entities that he postulates:

Hitherto I have laid down the definitions of such words as are less known, and explained the sense in which I would have them to be understood in the following discourse. I do not define time, space, place, and motion, as being well known to all. Only I must observe, that the common people conceive those quantities under no other notions but from the relation they bear to sensible objects. And thence arise certain prejudices, for the removing of which it will be convenient to distinguish them into absolute and relative, true and apparent, mathematical and common.

I. Absolute, true, and mathematical time, of itself, and from its own nature, flows equably without relation to anything external, and by another name is called duration: relative, apparent, and common time, is some sensible and external (whether accurate or unequable) measure of duration by the means of motion, which is commonly used instead of true time; such as an hour, a day, a month, a year.

II. Absolute space, in its own nature, without relation to anything external, remains always similar and immovable. Relative space is some movable dimension or measure of the absolute spaces; which our senses determine by its position to bodies; and which is commonly taken for immovable space; such is the dimension of a subterraneous, an aerial, or celestial space, determined by its position in respect of the earth. Absolute and relative spaces are the same in figure and magnitude; but they do not remain always numerically the same. For if the earth, for instance, moves, a space of our air, which relatively and in respect of the earth, remains always the same, will at one time be one part of the absolute space into which the air passes; at another time it will be another part of the same, and so, absolutely understood, it will be continually changed.

III. Place is the part of space which a body takes up, and is according to the space either absolute or relative. . . .

IV. Absolute motion is the translation of a body from one absolute place into another; and relative motion the translation from one relative place into another. Thus in a ship under sail, the relative place of a body is that part of the ship which the body possesses; or that part of the cavity which the body fills, and which therefore moves together with the ship: and relative rest is the continuance of the body in the same part of the ship, or of its cavity. But real, absolute rest is the continuance of the body in the same part of that immovable space, in which the ship itself, its cavity, and all that it contains, is moved. Wherefore, if the earth is really at rest, the body, which relatively rests in the ship, will really and absolutely move with the same velocity that the ship has on the earth. But if the earth also moves, the true and absolute motion of the body will arise, partly from the true motion of the earth, in immovable space, partly from the relative motion of the ship on the earth. . . .[5]

Newton distinguishes the "absolute, true, and mathematical" notions of space, time, place, and motion from their "relative, apparent, and common" counterparts. The crux of the problem is that while Newton's Laws of Motion are framed in terms of the absolute notions, these do not fall under our immediate observation. When we try to observe the motion of an object, all we can directly see is its relative motion: the change in its position with respect to other visible objects, with the rate of change being measured by the visible motion of clocks or other instruments for telling time. But if the absolute motion of an object is imperceptible because absolute space and time are imperceptible, how can the postulation of such entities have any relevance to empirical science?

Newton devotes the rest of the Scholium to answering this question:

[5] Newton (1934), vol. 1, pp. 6–7.

But because the parts of space cannot be seen, or distinguished from one another by our senses, therefore in their stead we use sensible measures of them. For from the positions and distances of things from any body considered as immovable, we define all places; and then with respect to such places, we estimate all motions, considering bodies as transferred from some of those places into others. And so, instead of absolute places and motions, we use relative ones; and that without any inconvenience in common affairs; but in philosophical disquisitions, we ought to abstract from our senses, and consider things themselves, distinct from what are only sensible measures of them. For it may be that there is no body really at rest, to which the places and motions of others may be referred.

But we may distinguish rest and motion, absolute and relative, one from the other, by their properties, causes, and effects. . . .[6]

Newton produces powerful *empirical* evidence for the existence of absolute motion (and hence absolute space and time) using considerations of the causes of motion. For this argument, we will need to consider his Second Law.

But before turning to the Second Law, we should pause to reflect how deeply intuitive Newton's account of absolute space and time is, even though absolute space and time are not directly observable. It sounds as if Newton is postulating some weird, ghostly, unfamiliar entities, but most people conceive of the physical world in terms of absolute space and time. For example, craftsmen and scientists continually try to improve the design of timepieces, to produce clocks that are ever more accurate and precise. But what is it for a clock to be "accurate"? What we want is for the successive ticks of the clock to occur *at equal intervals of time*, or for the second hand of a watch to sweep out its circle *at a constant rate*. But "equal" or "constant" with respect to what? With respect to the passage of time itself, that is, with respect to absolute time. Our natural, intuitive view is that a certain amount

---

[6] Ibid., p. 8.

of absolute time elapses between the successive ticks of a clock, and the better and more accurate the clock is, the more similar these intervals are to one another. Swiss watchmakers, and designers of atomic clocks, are trying to get their devices to accurately measure something, and that something is not any sort of relative, observable time. Physics treats every observable physical motion—the rotation of the earth, the motion of the earth around the sun, and so on—as subject to disturbances and hence not automatically uniform. But the nonuniformity is not defined with respect to any observable motion. Clock design reflects this commitment: disturbing factors are eliminated or compensated for. This practice implicitly assumes some measure of time itself that provides the standard of uniformity.

Similarly, our everyday understanding of the world conceives of it in terms of absolute space. No one is puzzled upon hearing, for example, that the orbit of the earth is an ellipse with the sun at one focus. Any picture of the solar system in a science book will draw the orbits of the planets. But what, exactly, is this supposed to be a picture of? At any given moment, the earth is in some one place. The "orbit" is somehow a collection of all the places the earth occupies over the course of a year. But that implies that the places the earth occupies at different times are all parts of one common, three-dimensional space: absolute space. We can, with effort and careful thought, come to comprehend how the world could exist without any absolute space or absolute motion. But when we do so we not only reject Newton's theory, we reject common sense as well.

Newton did not appeal to common sense to justify his belief in absolute space and time: he appealed to experiment. Newton tried to prove the existence of absolute motion in the laboratory rather than by conceptual analysis. This is our next topic.

Evidence for Spatial and Temporal Structure

Newton's First Law, the so-called Law of Inertia, refers only to bodies that are subject to no external forces. It is tempting to say that Newton postulates that such bodies "continue in the same state of motion," but such a formulation would miss the revolutionary aspect of this law: the First Law specifies exactly *what counts* as "the same state of motion." For Aristotle, as we have seen, a piece of aether in uniform circular motion about the center of the universe is always in "the same state of motion," and so there would be no reason to seek out external causes in such a situation. In Aristotle's physics, external causes are responsible for unnatural motion, such as a rock moving upward instead of down. So for Aristotle, the falling of a stone that is initially at rest but unsupported requires no external cause, and the continued uniform rotation of the sphere of fixed stars requires no external cause: this what these sorts of matter do by nature.

It is interesting to note that prior to Newton, Galileo sought to specify the "inertial" motion of terrestrial objects—that is, the motion that they would display if subject to no forces—and he concluded that such motion would be *uniform circular* motion. He arrived at this conclusion from his experimental work with inclined planes. Galileo noticed that if a ball rolls down one inclined plane and then up another, the ball very nearly rises back up to the height at which it began. He concluded, rightly as it turns out, that absent friction and air resistance the ball would roll up the second plane until it reached exactly the height from which it began (figure 2). As the angle θ is made smaller and smaller, the ball will have to roll farther and farther up the second ramp

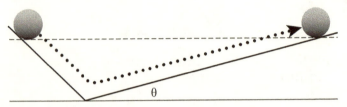

Fig. 2

before it reaches its original altitude. If the angle is set to 0, so the ball rolls without gaining any altitude at all, it would continue rolling forever (in the absence of friction) at a constant speed.[1]

What sort of motion is this, a motion at a constant speed that never gains or loses altitude? It is uniform *circular* motion about the center of the earth! Such a ball would roll around and around the earth forever, never speeding up or slowing down. The ball would certainly not continue forever in a straight line: a straight line, tangent to the earth's surface, leaves the earth and rises. A ball attempting to follow a straight path would slow down, eventually stop, and then roll back down to its starting point. Galileo concluded that the natural, inertial motion of all objects was like the natural motion of Aristotle's aether: uniform circular motion. For Galileo, this *circular inertia* helped explain how, for example, birds could keep up with the rotation of the earth. They did not need to exert a lot of effort, even though they would have to travel at hundreds of miles per hour in order not to be left behind: their circular inertia would carry them along without need for any force.

Newton's innovation is not merely the postulation of some inertial motion that a body tends to maintain when external force is absent but the particular claim that this is uniform motion in a straight line. But the First Law is clearly insufficient to account for the world around us: we never see bodies continuing in uniform motion in a straight line. The First Law needs to be supplemented by an account of the effect of external forces on the motion of a body, which is exactly what the Second Law does:

[1] Galileo (1967), pp. 145–149.

Law II: The change in motion is proportional to the impressed motive force and is made along the straight line on which the force is impressed.[2]

The Second Law is conceptually parasitic on the First Law: without the definition of a "state of motion" in the First Law, we could make no sense of "change in motion" in the Second. According to Aristotle and to Galileo, certain uniform circular motions are constant motions: they display no change of motion, and so they call for no special explanation. But according to Newton, a body in uniform circular motion is constantly changing its state of motion, and so it must be subject to some external force.

Newton's concept of "force," as employed in the Second Law, is not exactly the same as that used in modern physics. Newton's notion corresponds rather to what would now be called an "impulse," that is, the action of a force over some period of time. For Newton, the same "force" can be applied to a body by pushing harder over a shorter time or by pushing less hard over a longer period. If we push twice as long but with the same strength, then we subject a body to twice the force and will change its state of motion twice as much.

Let's consider how Newton's Second Law applies to uniform circular motion. Figure 3 indicates—in absolute space—the trajectory of a particle in uniform counterclockwise circular motion about point o. At exactly noon, let us say, the particle is at point a. The velocity of the particle is to the left along the tangent line T. According to the First Law of Motion, if there is no force on the particle after noon it would continue along T at a constant rate. So in order to continue along the circular path toward b, some force must be applied that will deflect the particle downward, toward o. Furthermore, if we want the particle to continue at a constant rate, without speeding up or slowing down, then no part of the force can be directed along with the present velocity (which would speed up the particle) or opposite to that direction (which would slow down the particle). So according to the Second Law, to keep the particle in uniform circular motion, there

[2] My translation.

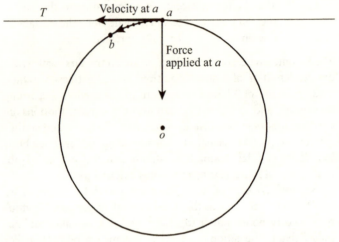

Fig. 3

must be a constant force impressed, and that force must always be directed exactly at the center of the circle. This is called a *centripetal* (center-seeking) force.

While Aristotle claimed that the moon, being made of aether, would naturally orbit the center of the universe (and hence the earth), Newton's First and Second Laws of Motion imply that the behavior of the moon requires the existence of a force directed at the earth, which keeps the moon from continuing on in a straight line. That force is gravity.

The association of forces with changes of absolute motion provides the key to Newton's strongest argument in favor of absolute motions and, by extension, in favor of absolute space and time. For even if the absolute motions of bodies are not immediately observable, sometimes the *forces* on those bodies are. And where there is a net force, there is, according to the Second Law, a change of absolute motion.

In order to illustrate this connection, Newton proposes an experiment so simple and familiar that there is no need to actually carry it out. Fill a bucket with water, hang it from the ceiling by a rope, and strongly twist the rope. Upon being let go, rope begins to unwind, and the experiment proceeds in four stages:

> [T]he surface of the water will at first be plain [i.e., flat], as before the vessel began to move; but after that the vessel, by gradually communicating its motion to the water, will make it begin sensibly to revolve, and recede by little and little from the middle, and ascend to the sides of the vessel, forming itself into a concave figure (as I have experienced), and the swifter the motion becomes, the higher will the water rise, till at last, performing its revolutions in the same times with vessel, it becomes relatively at rest in it. This ascent of the water shows its endeavor to recede from the axis of its motion; and the true and absolute circular motion of the water, which is here directly contrary to the relative, becomes known and may be measured by this endeavor. . . . [T]his endeavor does not depend on any translation of the water in respect of the ambient bodies, nor can true circular motion be defined by such translation. There is only one real, circular motion of any one revolving body, corresponding to only one power of endeavoring to recede from its axis of motion, as its proper and adequate effect; but relative motions, in one and the same body, are innumerable, according to the various relations it bears to external bodies, and, like other relations, are altogether destitute of any real effect, any otherwise than they may perhaps partake of that one only true motion.[3]

As familiar an occurrence as water sloshing up in a spinning bucket provides Newton the ammunition he needs to attack the idea that the relative motion of bodies is all the motion there is. Or, more precisely, it provides Newton with ammunition to attack the relationist account of motion advocated by Descartes and Aristotle. One obvious difficulty with a relationist theory of motion is that any body will have many distinct relative motions, since there are many other bodies to which it can be compared. But a physics of motion requires that there be one special motion of

---

a body, namely the motion that enters into the laws of physics. Aristotle and Descartes settled on the same idea: the important physical motion of a body is its motion with respect to the body that immediately encloses it.[4] Let's try to apply this account to the bucket experiment. At the beginning of the experiment, before the bucket is let go, the water is at rest relative to the bucket, and the surface of the water is flat. Later, when the water and bucket spin together, the water is again at rest with respect to the bucket (and hence, according to Descartes and Aristotle, truly at rest), but the concave surface indicates that the situation is physically different. We say that this difference is due to the water not spinning in the earlier stage and spinning in the later stage, but this cannot be motion *relative to the bucket* or relative to the *body that encloses it*.

The relativist can try to respond that the relevant motion, the motion that accounts for the behavior of the water, is not motion relative to the bucket but is relative motion all the same. The physically effective spinning is perhaps spinning *with respect to the room*, or *with respect to the earth*, or *with respect to the fixed stars*. But at the end of the Scholium, Newton presents the purest possible example of observable effects of absolute motion in the absence of any relative motions.

> For instance, if two globes, kept at a given distance one from the other by means of a cord that connects them, were revolved about their common center of gravity, we might, from the tension of the cord, discover the endeavor of the globes to recede from the axis of their motion, and from thence we might compute the quantity of their circular motions. . . . And thus we might find both the quantity and the determination of this circular motion, even in an immense vacuum, where there was nothing external or sensible with which the globes could be compared.[5]

---

[4] Aristotle's statement of this doctrine of place—and hence change of place—may be found in chapter 4 of book IV of the *Physics* (212ª20). Descartes makes the same claim in the *Principles of Philosophy*, part 2, sec. 25.

[5] Newton (1934), vol. 1, p. 12.

The observable tension in the cord, which produces a centripetal force, testifies to the absolute rotation of the spheres even when there is no relative motion in the entire universe: all material bodies (the globes and the cord) maintain a constant position with respect to one another.

Newton's bucket experiment, for all its simplicity, remains one of the most powerful and compelling experiments in the history of physics. The behavior of the water in the bucket, or the tension in the cord connecting the globes, is an observable fact that requires an explanation. The natural explanation occurs to us: the water ascends the sides of the bucket and the cord displays tension because the system is *spinning*. But spinning is a sort of motion, so we must ask: spinning *with respect to what*? The relevant motion is not motion with respect to the immediate surroundings. And if we accept the thought experiment with the globes in an otherwise empty space, the relevant motion cannot be motion with respect to any material body. Newton concludes that the motion must be motion with respect to absolute space: the spinning bodies successively occupy different locations in space itself. In this way, absolute motions are connected to forces and hence to observable effects.[6]

Nowadays, almost no one believes in absolute space persisting through time. Therefore, in Newton's sense, no one believes in absolute motion. Nonetheless, even in the Theory of Relativity, Newton's bucket experiment still proves that the *relative motion of bodies* does not explain the phenomena Newton discusses. Even in Relativity, it remains true that a pair of yoked globes, in an otherwise empty universe, could display a tension in the cord that connects them, and that tension would be an indication that the globes are not, in some absolute sense, following their natural, unencumbered trajectories. Even in Relativity, there is an *absolute* fact about whether the globes are rotating about their common center of gravity or not.[7]

---

[6] A more detailed and nuanced analysis of Newton's Scholium can be found in Rynasiewicz (1995).

[7] It is perhaps worthwhile to point out that the description of Newton's argument for absolute space offered in some of the philosophical literature is not

It is not obvious how to retain a notion of absolute rotation, rotation "relative to space itself," without Newton's absolute space and time. The conceptual resources needed to understand how to decouple the heart of Newton's dynamics from absolute space and time were not developed until centuries after Newton's death. Even today, few physics textbooks explain the situation clearly. Indeed, in order to understand how we can have absolute rotation without absolute space and time, we need to reconceptualize the geometry of space and time, welding them together into a single entity. But before we take that step, let's pause for some brief remarks on how mathematics is used in expounding Newton's theory.

ARITHMETIC, GEOMETRY, AND COORDINATES

I have presented Newton's First and Second Laws of Motion without recourse to a single mathematical equation.[8] It is said that nontechnical presentations of physics ought to avoid equations for fear of intimidating the reader, but that is not the reason I have not used the standard equations, such as $F = mA$. I have not used

---

strictly accurate. In his classic *Space, Time, and Spacetime* (1977), Lawrence Sklar writes that "accelerations give rise to *observable* forces," which he calls "inertial forces," and says that Newton postulates absolute motion to account for these forces. Among the "inertial" forces, according to Sklar, are "centrifugal forces throughout the astronomical universe (it is the same force that 'keeps the planets from falling into the sun')" (pp. 183–184). But Newton postulates no such thing as "inertial" forces: there are only physical forces, such as the force due to gravity. And accelerations do not cause forces: forces cause accelerations (changes in motion). There is no centrifugal (center-fleeing) force that "keeps the planets from falling into the sun": there is only a centri*petal* force that keeps the planets from *drifting away* from the sun. That force is the force of gravity.

In physics texts, "inertial forces" are correctly described as *fictitious* forces, i.e., there really are no such things. One can fooled into thinking there are such forces by improperly using a "non-inertial reference frame." What exactly *that* means is the subject of the next few sections.

[8] I have left aside the Third Law, which asserts that for every action of one body on another, there is an equal and opposite action of the second on the first, as it is not central to our concerns. One can argue that it follows from the first two laws: if it were violated, a system of two bodies could spontaneously accelerate.

equations because Newton did not use equations: he presented his theory geometrically and proved theorems in the *Principia* by the techniques of Euclidean geometry. This is not just a historical curiosity. Newton presented his physics geometrically because the subject matter—motion in space—is itself geometrical. To put it bluntly, the physical world is not composed of *numbers* or of entities for which the standard arithmetic operations are defined. The physical world does contain *physical magnitudes* that have a geometrical structure. Geometry is more directly connected to the physical world than is arithmetic.

For example, I have said that Newton presupposes the geometrical structure of absolute space to be E^3: the three-dimensional geometry described by Euclid's postulates. If that is correct, then the use of Euclidean geometry is a direct route to the derivation of important physical facts. But modern physics is never presented directly in terms of Euclidean geometry, and contemporary students might not easily follow Newton's proofs or construct similar proofs. Rather, modern physics is presented *algebraically*, in terms of *numbers* and *arithmetic equations*. This association of physics with arithmetic has become so deeply ingrained that it can be hard to even see. For example, a typical physics text will not refer to three-dimensional Euclidean space as E^3; it will refer to it instead as R^3. And "R^3" has a particular mathematical meaning: R^3 is the set of ordered triples of real numbers. An element of R^3 has the form (x, y, z), where each of the variables stands for a particular real number: $(1, 3.52, \sqrt{17})$ is an element of R^3. But $(1, 3.52, \sqrt{17})$ is clearly not an element of three-dimensional Euclidean space. Furthermore, R^3 has quite a bit of mathematical structure that has no analog in E^3. For example, given any element of R^3, we can ask whether the numbers in the first and second slots are the same or different. And given any two elements of R^3, we can ask whether they have the same or different numbers in their first slots. Given any pair of elements of R^3, (x_1, y_1, z_1) and (x_2, y_2, z_2), it is easy to define their *sum* as $(x_1+x_2, y_1+y_2, z_1+z_2)$, and easy to define what it means to multiply an element by a scalar c: $c \times (x_1, y_1, z_1) = (cx_1, cy_1, cz_1)$. But none of these operations have analogs for the points in E^3. Points in E^3 don't have "first slots" and "second slots"; they cannot be "added" in any well-defined way, and it makes no sense to ask what the result of multiplying a point by a scalar is. R^3 and E^3 are clearly different mathematical objects. How is it, then, that the two have come to be so easily confused?

The answer lies in the use of *coordinate systems*. A coordinate system on E^3 establishes a one-to-one correspondence between the elements of E^3 (geometrical points) and some sort of arithmetical object. We commonly use elements of R^3 as coordinates for points of E^3. Given such a coordinate system, the triple of real numbers $(1, 3.52, \sqrt{17})$ comes to *represent* a point in E^3. The advantage of using such arithmetical representations of geometrical objects is that it allows us to use *algebraic* methods to solve *geometrical* problems. By means of a coordinate system, problems of geometry can be converted into questions of arithmetic, and arithmetical methods can be used to solve them.

The mathematical power of this translation of geometry into arithmetic cannot be overstated. But with it comes an accompanying danger: the distinction between the object of study (a geometrical space) and a *representation* of that object (say, ordered n-tuples of numbers) can become obscured. This is not an idle worry: Einstein reports in his autobiographical notes concerning the long incubation time for General Relativity:

> Why were another seven years required for the construction of the general theory of relativity? The main reason lies in the fact that it is not so easy to free oneself from the idea that coordinates must have an immediate metrical meaning.[9]

Even Einstein had a hard time distinguishing those features of coordinates that carry some geometrical or physical significance from those that don't. In order to appreciate how coordinates can encode geometrical structure, let's consider coordinatizing the two-dimensional Euclidean plane E^2.

Everyone is familiar with Cartesian coordinates of the Euclidean plane: every point in the plane receives an x-coordinate and a y-coordinate, each of these being a real number. So each point becomes associated with an element (x,y) of R^2. Now,

[9] Einstein (1982), p. 67.

already, we might ask why *two* real numbers are being used instead of just *one*. After all, it is possible to code up the same information that is contained in a pair of real numbers in just a single number. For example, we can interleave the digits in the decimal expansions of the real numbers: the pair of real numbers (.12341234 . . . , .88888 . . .) would get mapped to the single real number .1828384818283848. . . . Such a system is, in a way, more efficient than the usual scheme: instead of needing a pair of numbers to indicate a particular point in space, a single number will suffice.

But using such a method has serious drawbacks. If we were to associate points in the plane with numbers by this scheme, *the continuous motion of a point in the plane would be represented by discontinuous changes in the coordinates of the point*: nearby points in the plane would not always have nearby coordinate values. Here's an example. Start with the standard Cartesian coordinates on the plane, then convert to a single coordinate by interleaving the digits as above. Now consider a point that starts at the origin (0,0) of the Cartesian coordinate system and moves smoothly along the x-axis to (.9,0). In this motion, the y-coordinate is always 0. Hence, in terms of our single coordinate the motion starts at .00000 . . . and ends at .90000, but never passes through the point labeled .01000. . . . The continuous motion on E^2 is represented by a discontinuous change in the coordinate. But if we use a pair of Cartesian coordinates, this problem is solved: as a point moves continuously in the space, its coordinates each change continuously.

This feature of a well-adapted coordinate system is so basic, and so obvious, that no one would ever recommended anything like our "interleaving" procedure. But for all its obviousness, it deserves some explicit recognition. This basic requirement for any acceptable coordinate system is that the topology of the coordinates should mesh properly with the topology of the space being coordinatized. The coordinate functions, whose domain is the space and whose range is (in this case) the real numbers, are required to be continuous functions. In the case of Cartesian coordinates on E^2, there are two such functions: an x-coordinate function $X(p)$ and a y-coordinate function $Y(p)$. For every point p in E^2,

27

these functions define a pair of numbers which are the coordinates of the given point $(X(p), Y(p))$. If these coordinate functions are everywhere continuous, then we can recover the topology of the space itself from the topology of the coordinates. The coordinates come to represent, in the most straightforward way possible, the geometrical structure of the space being coordinatized.

Once we see how a well-chosen coordinate system can be adapted to the topology of the space being coordinatized, it naturally occurs to ask whether the other geometrical features of the space can be reflected in the coordinates in a similarly transparent way. We can ask this about the differential structure, the affine structure, and the metrical structure of the space.

Given a differential structure, we can naturally demand that the coordinate curves be not just continuous but also smooth: they should not have sharp corners. This is usually presumed without comment.

In the case of Cartesian coordinates on E^2, the connection between the geometry of the space and the algebra of the coordinates is particularly clear. Consider the affine structure first. Let A, B, C, and D be any four real numbers, with A and C not both 0. Then the set of points with Cartesian coordinates $(As + B, Cs + D)$ form a complete straight line, with s ranging over the real numbers. For example, the x-axis corresponds to the choices $A = 1$, $B = C = D = 0$; the y-axis corresponds to $C = 1$, $A = B = D = 0$; and the $x = y$ diagonal corresponds to $A = C = 1$, $B = D = 0$. Conversely, every straight line in E^2 corresponds to at least one (in fact, infinitely many) choice for A, B, C, and D. So straight lines in E^2 correspond to certain *linear equations* for the coordinates.

What about the metrical structure of E^2? Metrical structure concerns the distances between points. And the distance between two points in E^2 is the length of the straight line that connects them. So the metric determines, in the first place, *ratios of the lengths of straight lines*. If all such ratios are determined, then the set of circles is also determined: a circle around point p is the set of all points that are equidistant from p, that is, whose distance from p is in the ratio 1:1.

The metrical structure of E^2 is particularly simple to express in terms of Cartesian coordinates. A *metric function* is a function

$D(u,v)$ from pairs of points in a space to the real numbers that satisfies three conditions:

1: $D(u,v) \geq 0$, with $D(u,v) = 0$ if and only if $u = v$ (Positive Definiteness)
2: $D(u,v) = D(v,u)$ (Symmetry)
3: $D(u,v) + D(v,w) \geq D(u,w)$ (Triangle Inequality)

$D(p,q)$ is sometimes called the *distance* from p to q, but this terminology is a bit misleading. $D(x,y)$ is a function from points in the space to the real numbers. But if we were to ask for the distance from point p in E^2 to point q, the correct answer cannot possibly be a real number, such as 4; 4 *what*? When we specify a distance using a real number, we always have to employ a unit or a scale. So the particular value of the function $D(u,v)$ has no geometrical significance.

What does have geometrical significance are *ratios* between the values of the metric function. If $D(p,q) = 2$ and $D(r,s) = 4$, then the ratio of the length of the straight line $\overline{pq}$ to the length of $\overline{rs}$ is 1:2, that is, $\overline{pq}$ is twice as long as $\overline{rs}$. So two metric *functions* that differ from one another by a constant scale factor represent the same metrical *structure* in the space. The freedom to choose any of these different functions to represent the same geometry is called a *gauge freedom*. We encounter many such gauge freedoms when we use numbers to represent geometrical structure, and it is important to develop a sense of when this happens.

In the case of E^2 with Cartesian coordinates, a metric function is specified by means of the coordinates of the points, using the Pythagorean formula:

$$D(p,q) = \sqrt{\left(X(p) - X(q)\right)^2 + \left(Y(p) - Y(q)\right)^2} \text{ (Equation 1)}.$$

The arguments of this function are points p and q in E^2. These points have no arithmetic properties, so it makes no sense to subtract them or square the result. But the coordinates of these points, $X(p)$, $Y(p)$, $X(q)$, and $Y(q)$ are real numbers: they can be subject to the usual arithmetic operations.

The complete geometrical structure of E^2—its topology, affine structure, and metric—can be easily recovered from a set of Cartesian coordinates. And, more importantly, any space can be identified as E^2 by the criterion that it *admits of* Cartesian coordinates. That is, a space is E^2 if and only if there exists a pair of functions X and Y from the space to the real numbers such that (1) X and Y are both continuous, (2) the complete straight lines correspond to sets of points whose coordinates are $(As + B, Cs + D)$ for some choice of A, B, C, and D, and (3) the ratios of lengths of straight lines are the ratios of the values of the function given by equation 1. A space has to have a certain geometrical form to admit of such a coordinatization so we can use the existence of such coordinate systems as an indirect means to specify the geometry of a space.

Once a coordinate system has been chosen for a space, the dynamical laws (i.e., the laws that specify the trajectories of objects in the space) can be stated in algebraic form *relative to those coordinates*. This is what is commonly done in contemporary physics books when presenting Newtonian mechanics. For example, Herbert Goldstein's text *Classical Mechanics* presents Newton's theory as follows:

> Let **r** be the radius vector of a particle from some given origin and **v** its vector velocity:
>
> $$\mathbf{v} = \frac{d\mathbf{r}}{dt}.$$
>
> The *linear momentum* **p** of the particle is defined as the product of the particle mass and its velocity:
>
> $$\mathbf{p} = m\mathbf{v}.$$
>
> In consequence of interactions with external objects and fields the particle may experience forces of various types, e.g. gravitational or electrodynamic; the vector sum of these forces exerted on the particle is the total force **F**. The mechanics of the particle is contained in *Newton's Second Law of Motion*, which states that there exist frames of reference in which the motion of the particle is described by the differential equation
>
> $$\mathbf{F} = \frac{d\mathbf{p}}{dt}. \text{[10]}$$

---

[10] Goldstein (1981), pp. 1–2.

Goldstein's description of Newton's theory has little in common with Newton's own account. Goldstein's very first sentence invokes a vector "from some given origin," while Newton makes no mention of an "origin." Already, Goldstein is presupposing (without mentioning it) some coordinatization of the space, with the "origin" being the point with coordinates (0,0,0). Furthermore, all of the symbols Goldstein uses—viz., $\mathbf{r}$, $\mathbf{v}$, $\mathbf{p}$, $\mathbf{F}$, m, and t—are taken without comment to be numbers or to have algebraic properties: $\mathbf{v}$ can be *multiplied* by m, and the derivatives (which are defined by limits of arithmetic operations) are supposed to exist. Most strikingly, according to Goldstein, Newton's Second Law of Motion states something about the existence of "frames of reference." But, as we have seen, Newton states his Second Law without mention of any such thing: indeed, he would have had no idea what a "frame of reference" *was*. It is also striking that Goldstein somehow jumps to the Second Law without so much as a mention of the First Law, but as we have seen, Newton's Second Law has to *presuppose* the First Law in order even to have any content.

In short, modern physics has become so thoroughly arithmetized, so pervasively dependent on numbers introduced by means of coordinate systems, that we cannot readily recognize Newton's original theory in a modern presentation. The modern presentation is much more efficient when it comes to solving problems: use of the calculus eliminates the need for Newton's diagrams and geometrical constructions. But it has become correspondingly obscure exactly what the structure of the theory is and, in particular, what the theory is claiming about the world. It is, for example, sometimes said that Newton's Laws of Motion only hold in *inertial frames of reference*. And if we ask further what exactly an inertial frame of reference is, it is tempting to reply that it is a frame of reference in which Newton's Laws hold. Such a tight definitional circle ought to make us uncomfortable, but there is an acceptable understanding of the procedure. Just as the intrinsic geometrical structure of E^3 entails that Cartesian coordinate systems exist, so the intrinsic geometrical structure of space and time according to Newton entails that special sets of coordinates exist. In the most convenient coordinatizations of Newtonian space and time, the acceleration of a trajectory through time is proportional to the second derivative of the spatial coordinates with respect to the time coordinate. For example, the change in state of motion in a given direction (let us call it the x-direction), which for Newton is a quantity that exists independently of all coordinates, becomes proportional to d^2x/dt^2, where x and t are coordinates in this special system. In such an approach, the existence of such convenient coordinates is a derivative matter: they follow from the space-time structure itself. Newton's Laws (as Newton formulated them) describe the motion of bodies, in a space with the geometrical structure of E^3 that persists through time. No mention of an "inertial frame of reference," or of *any* "frame of reference" is required.[11]

Coordinate systems themselves can be characterized by reference to the geometrical structure of the space being coordinatized. For example, Cartesian coordinates on a Euclidean space are said to be *rectilinear* and *orthogonal*. What this means is that the *coordinate curves* are straight lines that meet at right angles. A coordinate curve is the set of points in the space we get if all but one of the coordinates is held fixed and the remaining coordinate is allowed to vary. So, for example, take some Cartesian coordinates (x, y) on E^2. If we fix the x coordinate as 2 and let the y coordinate vary, we get all the points with $x = 2$. This forms a "vertical" line on the plane (figure 4). The x-coordinate curves in a Cartesian coordinate system are the "horizontal" straight lines, and the y-coordinate curves are the "vertical" straight lines. Any pair of coordinate curves that meet form a right angle where they meet.

Clearly, the characterization of a coordinate system as "rectilinear" and "orthogonal" depends upon the space itself, independently of all coordinates, having a certain geometrical structure. If the space does not have an affine structure, then the coordinate curves cannot be classified as either straight or curved; if it does not have a metrical structure, then the angle at which curves meet cannot be classified as right or acute or obtuse. So this sort of taxonomy of coordinate systems presupposes a certain geometrical structure in the space itself.

---

[11] More detailed discussion of the history of the concept of an inertial frame can be found in DiSalle (1991); Barbour (2001), chap. 12; and Brown (2006), chap. 2.

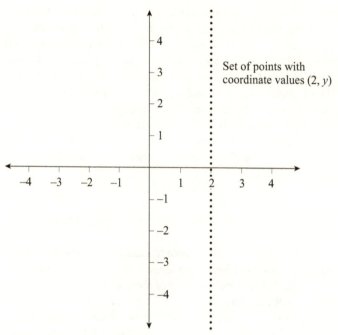

Set of points with coordinate values $(2, y)$

Fig. 4

Similarly, characterizing a coordinate system used in physics as "inertial" or "non-inertial" also presupposes that physical space and time have some sort of objective geometrical structure. But the exact nature of that geometrical structure is not so evident. We have seen the features that Newton himself attributes to absolute space: it has a three-dimensional Euclidean geometry, and the individual points of absolute space persist through time, retaining their geometrical relations. It turns out that this is more structure than is required for Newton's dynamics: his Laws of Motion can be formulated without reference to absolute space and time. It takes some patient groundwork, though, before we can see how to do this.

Having made this foray into coordinate systems and arithmetical representations of geometrical structure, I will try, insofar as possible, to present the physical ideas without using coordinates or numbers. The most effective means for doing this is graphical: diagrams can be used as geometrical representations of geometrical facts, and they can provide a quick intuitive grasp of the fundamental ideas. Interpreting these diagrams requires some care because the diagrams, as spatial objects, often have geometrical features that do not correspond to anything in the space represented. So the reader should study the diagrams carefully but also bear in mind remarks about how the diagrams should be interpreted.

We will now pick the up the thread of our historical survey where we left off.

THE SYMMETRIES OF SPACE AND THE LEIBNIZ-CLARKE DEBATE

We now have an account of some fundamental geometrical structure that a space can have—topological, affine, and metrical—and have seen that Newton's account of absolute motion presupposes a space with all three features whose individual points persist through absolute time. Absolute time itself has a topology, a metrical structure, and also, unlike space, a directionality. Explication of Newton's Laws of Motion, however, does not make any essential use of the Euclidean structure of the space: although Newton presumes space to be E^3, nothing in his dynamics requires this. But the particular choice of E^3 gave rise to a set of objections to Newton's theory.

Three-dimensional Euclidean space displays a high degree of symmetry. These symmetries manifest themselves in isometries of the space: mappings of the space onto itself that preserve distances among the points. The isometry depicted in figure 1 illustrates two of these symmetries: translational symmetry and rotational symmetry. That is, in order to generate the bottom image in figure 1 from the middle image, we can imagine first moving all the points an equal distance downward on the page (translation) and the turning the whole thing about an axis (rotation). All of the parts of the diagram bear the same geometrical relations to one another after the translation and rotation have been carried out.

A space that has a translational symmetry is called *homogeneous*, and one that has a rotational symmetry about a point is called *isotropic*. In a homogeneous space, all the locations "look the same": you can move everything over in the space and retain the same geometrical relations among them. If a space is isotropic at a point, then all the directions at that point "look the same": everything can be reoriented by rotation about the point and still retain all geometrical relations.

Aristotle's spherical universe is not homogeneous: it is impossible to translate everything in any direction because the boundary of the space does not allow it. Aristotle's physics reflects this inhomogeneity. A stone moved away from the center of the universe would not behave the same as it did at the center: from its new location it would spontaneously move back toward the center. Aristotle's universe is isotropic at the center but not at any other point: if you are not at the center of the universe, then there is a unique "downward" direction that points to the center. But E^3 has complete translational and rotational symmetry. Every point can be shifted over a fixed distance in any given direction without changing the geometrical relations among the points, and every point can be rotated about any axis without changing the geometrical relations among the points. These symmetries are implicitly assumed in Euclidean geometry: we should be able to draw any Euclidean diagram and get the same result no matter where in the world the paper is located or how the paper is oriented. Otherwise, the instructions for drawing a Euclidean diagram would have to specify where it must be drawn.

As a consequence of these symmetries of E^3, it is possible for matter to be distributed in different ways in E^3 but to retain all the same *relative* locations. Take any distribution of matter and move it all over a fixed distance, or rotate it all the same amount around an axis, and the relative positions of objects will not change. If the matter distribution itself does not have the same symmetry, then the new distribution is distinct from the old one, but all relative positions remain the same.

In a famous indirect exchange of letters, Gottfried Leibniz argued with Samuel Clarke over the acceptability of Newton's physics. The initial topic was whether Newton's physics would undermine religious belief. In the course of the discussion, Leibniz inserts an advertisement for his own philosophy. He claims to have raised metaphysics to a demonstrative science by means of a single principle: the Principle of Sufficient Reason (PSR):

> Now, by that single principle, viz. that there ought to be a sufficient reason why things should be so, and not otherwise, one may demonstrate the being of a God, and all the other parts of metaphysics or natural theology; and even, in some measure, those principles of natural philosophy, that are independent upon mathematics: I mean the dynamical principles, or the principles of force.[12]

Clarke, seeming to agree with Leibniz but intent upon illustrating how the will of God could be the ultimate cause of things, happened upon a fateful example:

> 'Tis very true, that nothing is, without a sufficient reason why it is, and why it is thus rather than otherwise. And, therefore, where there is no cause, there can be no effect. But this sufficient cause is oft-times no other, than the mere will of God. For instance: why this particular system of matter, should be created in one particular place, and that in another particular place; when, (all place being absolutely indifferent to all matter,) it would have been exactly the same thing *vice versa*, supposing the two systems (or the particles) of matter to be alike; there can be no other reason but the mere will of God.[13]

Thus began one of the most significant, but also most convoluted and confused, arguments in the history of philosophy.

Since Clarke appears to be endorsing Leibniz's PSR, it seems odd that this passage could set off a firestorm of argument. But, as Leibniz saw, Clarke's example does not really respect the PSR. Clarke says that when creating matter, God could place the entire system of matter anyplace at all in absolute space, "all place being absolutely indifferent to all matter." Suppose that God decides to

---

[12] Second Paper, ¶ 1. All translations are from Alexander (1956).
[13] Second Reply, ¶ 1.

create, in a completely empty two-dimensional Euclidean space, matter in the shape of circle inside an isosceles triangle. As figure 1 illustrates, there are different ways that God can do this: the symmetry of Euclidean space ensures that such a system of matter would fit in any place and with any orientation. So according to Clarke, God must make a choice among all of these distinct possible ways that absolute space could contain matter. Clarke wants to insist that "the mere will of God" must be the determining factor because there is nothing that makes any of these possible distributions of matter superior to any other.

Leibniz picks up on this immediately:

> The author grants me this important principle; that nothing happens without a sufficient reason, why it should be so rather than otherwise. But he grants it only in words, and in reality denies it. Which shows that he does not fully perceive the strength of it. And therefore he makes use of an instance, which exactly falls in with one of my demonstrations against real absolute space, which is an idol of some modern Englishmen. . . .
>
> I say then, that if space is an absolute being, there would be something for which it would be impossible there should be a sufficient reason. Which is against my axiom. And I prove it thus. Space is something absolutely uniform; and without the things placed in it, one point in space does not absolutely differ in any respect whatsoever from another point of space. Now from hence it follows, (supposing space to be something in itself, beside the order of bodies among themselves,) that 'tis impossible that there should be a reason why God, preserving the same situation of bodies among themselves, should have placed them in space after one particular manner, and not otherwise; why everything was not placed the quite contrary way, for instance, by changing East into West.[14]

Let's call this the *PSR argument*, since it relies on the Principle of Sufficient Reason. The argument also uses some other

[14] Third Paper, ¶ 5.

premises. The premise that has received the most attention states that the existence and geometrical structure of absolute space is independent of all matter. Thus, it is possible for space to exist completely devoid of material objects: a perfect vacuum. A vacuum is not nothing: it is empty *space* and, according to Newton, such an empty space actually existed for an infinite period of time before God created any matter in it.

This brings us to the second premise of Leibniz's argument: God's creative act. Leibniz argues that even if God wanted to create a material world in an empty absolute space, He would be stymied: unable to find a grounds for locating the world one place rather than another, God would be prevented by the PSR from acting at all. This strikes Clarke as an impious denial of God's omnipotence, but Leibniz sticks to his guns: "When two things which cannot both be together, are equally good and neither in themselves, nor by their combination with other things, has the one any advantage over the other; God will produce neither of them."[15] Lacking any grounds for preferring one of two incompatible options, God *cannot* create either of them. Since the various locations of matter in absolute space would all be equally good, God could not will the existence of any of them, no matter how good they might be.

The involvement of God in the PSR argument threatens to make it irrelevant to anyone who does not accept a theological account of creation. What, for example, if *no one* created the material world, if it always existed? And there is a perfectly good sufficient reason for its location at any moment of time, namely, its location at previous times and the laws of physics. It is not entirely clear whether this sort of explanation would solve Leibniz's problem: even though the location of an object at any time is accounted for by its location at previous times, there is no reason why this particular distribution of matter through all time exists instead of a different one (e.g., one in which, for all time, the matter was "turned the other way" in absolute space).

Another premise of the PSR argument is the symmetry of absolute space. The claim that "one point in space does not

[15] Fourth Paper, ¶ 19.

absolutely differ in any respect whatsoever from another point of space" asserts the homogeneity of absolute space. In an Aristotelian spherical space, no such homogeneity exists, so there could be a reason for locating an object at the center of the space rather than at the periphery. Of course, Aristotelian space is isotropic about the center, so there would be no grounds for *orienting* matter one way or another, supposing such an orientation makes a difference. (If the matter itself formed perfect spheres, all centered at the center of the universe, then no issue of orientation would arise.) The appeal to these symmetries in the PSR argument is never challenged because both Leibniz and Clarke presume space to be E^3. But if we abandon that presumption and postulate a space that lacks these symmetries, the PSR argument cannot get off the ground. As we will see when we get to General Relativity, this is not idle speculation.

And, of course, the primary assumption of the PSR argument is the PSR itself. We can simply reject the PSR and accept that there are brute contingent facts about the world, facts that could have been different and for which there is no predetermining cause or explanation. Since, as Leibniz notes, Clarke avows that he accepts the PSR, this response to Leibniz is not pressed.

It should be noted just how strong a principle the PSR in full generality is. One application of the PSR is to dynamical laws: the principle is generally taken to be incompatible with dynamical indeterminism, that is, with the possibility that two physical systems in the same initial state and same environment could evolve differently. The history of quantum physics shows that physicists are not wedded to the PSR in this guise: they have been willing to take fundamentally indeterministic dynamical laws as serious physical proposals. But the PSR has much more radical consequences, as Leibniz saw. If there must be a sufficient reason why things are one way rather than any other, then there must be a reason why this exact universe, in all its particular detail, exists rather than any other physical possibility. That is, there must be a sufficient reason for the *initial state* of the universe, not just a sufficient reason for every alteration of physical state. Leibniz followed his principle to its logical conclusion: there must some intrinsic feature of the universe that explains why it, rather than

39

any other possible universe, exists. And for Leibniz, that reason is because this is the best of all possible worlds. Few of us would be willing to follow the PSR this far, but once we reject the principle, Leibniz's argument cannot get started.

If Leibniz had only given the PSR argument in his dispute with Clarke, it would not have been too hard to track down the premises of the argument and hold them up for examination. Unfortunately, Leibniz also employs a different argument, from completely different premises, in his attack on Newton. It requires some disciplined attention to keep these two arguments apart. It does not help matters that Leibniz introduces this second argument while summarizing his first and does not acknowledge that entirely new considerations are in play. The passage cited above continues:

> But if space is nothing else, but that order or relation; and is nothing at all without bodies, but the possibility of placing them; then those two states, the one such as it now is, the other supposed to be quite the contrary way, would not at all differ from one another. Their difference therefore is only to be found in our chimerical supposition of the reality of space in itself. But in truth the one would exactly be the same thing as the other, they being absolutely indiscernible; and consequently there is no room to enquire after a reason of the preference of the one to the other.[16]

The first part of this passage makes reference to Leibniz's relationist account of space, which we will return to presently. But the last sentence introduces a new consideration, based not on the PSR but on Leibniz's Principle of Identity of Indiscernibles (PII). This principle is distinct from the PSR, so arguments based on one must be carefully distinguished from arguments based on the other.

The PII argument, as formulated by Leibniz, runs as follows: if absolute space exists as Newton asserts, then the situation that obtains when matter is deployed in some configuration in space is different from the situation when it is deployed somewhere else in

[16] Third Paper, ¶ 5.

space, even if those two situations are related by an isometry. Picturesquely, we can imagine picking up all the matter and plunking it down somewhere else in absolute space. But these distinct configurations of matter in absolute space would "look exactly the same" in an obvious sense. No *qualitative* description could distinguish one from the other. In this sense, the two situations would be (qualitatively) indiscernible from one another. But according to the PII, there cannot be two distinct things (even two *merely possible* situations) that are indiscernible from one another.[17] Speaking loosely, we say that any two indiscernible things are identical, but properly speaking the claim is that there just can't be two indiscernible things. Since Newton's theory is committed to the possibility of distinct but indiscernible possible situations, it must contain a metaphysical error. That error is the postulation of absolute space.

It should first be noted that Leibniz's use of situations where matter is moved around in absolute space, producing different possibilities, is more complicated than necessary. Consider Newtonian absolute space completely void of matter, that is, consider a Newtonian vacuum. Empty E^3 is postulated to contain infinitely many distinct spatial locations, or points, all of which are qualitatively exactly alike. The homogeneity and isotropy of E^3 ensure that no qualitative description could apply to one point of absolute space but not to another. Since these distinct points of absolute space are supposed to be qualitatively identical, the PII implies that there cannot really be more than one of them, but E^3 contains more than one point. So the postulate of empty absolute space that is E^3 already violates the PII, quite independently of any further considerations about matter.

---

[17] Leibniz is not consistent about the exact status of the PII. In the passage cited above, it is clear that it is supposed to apply to the two (supposedly distinct) possibilities for locating matter in the universe and to have the consequence that they cannot be distinct possibilities. In Leibniz's Fifth Paper, ¶ 21, the principle is clearly weaker: there Leibniz says that distinct indiscernible objects, even in the same world, are possible, but that God will not produce them on account of the PSR. If Leibniz derives the PII from the PSR, then it falls when we reject the PSR, so only the independent justifications of the PII need concern us. An analysis of Leibniz's arguments in the *Correspondence* can be found in Rodriguez-Pereyra (1999).

The PII, if accepted, would also rule out most accounts of matter, independently of considerations about space. Leibniz himself thinks that the PII is inconsistent with the atomic theory of matter, since individual atoms are supposed to be qualitatively identical. This is one of the motivations of Leibniz's theory of monads, according to which the ultimate constituents of the world are unextended thinking entities, each of which represents to itself the entire universe from a unique perspective. It would follow that every monad is qualitatively different, with respect to the content of its representations, from every other one. Modern physics, of course, rejects this sort of constraint: two hydrogen atoms in the ground state are thought to be completely qualitatively identical.

So the main question about the PII argument is why we should accept the PII in the first place. Leibniz offers this brief justification in his Fourth Letter:

> To suppose two things indiscernible, is to suppose the same thing under two names.[18]

The most credible explanation of this claim is that Leibniz is presupposing a naïve version of an empiricist theory of ideas. On this view, a concept is something like a picture in the mind, which refers to whatever the picture depicts. But two qualitatively identical *pictures* depict the very same thing. So if we believe that we can even *imagine* two distinct but qualitatively indiscernible situations, we are confused: what is in the mind is the very same idea, with the very same reference, being used twice.

It is fair to say that this account of ideas, and how they refer to things, is now recognized to be untenable, and so Leibniz's justification for the PII has no force. There are, however, other sorts of worry that deserve some attention.

One objection to postulating indiscernible objects is more epistemological than metaphysical or semantic: if there are two indiscernible objects, then (almost by definition) we couldn't tell which was which. But then there would be certain questions that no observation could possibly answer. Postulating indiscernibles

[18] Fourth Paper, ¶ 6.

would entail a certain absolute limitation on what we can, even in principle, know.

As simple as this claim seems, it is actually rather hard to pin down. When we try to apply it to locations in absolute space, all of which are supposed to be qualitatively identical, the conclusion is that "we couldn't know by observation where, in absolute space, we are." But on reflection, there doesn't seem to be any way to articulate exactly what it is we couldn't know. What question is it, exactly, whose answer we cannot provide? *Where are we?* Well, what are the possible *answers* to this question that we are unable to decide between? As we will see in the next section, Newton's theory is committed to there being well-posed questions with different possible answers that cannot be settled by observations. But "Where in absolute space are we?" is not such a question.

And even if it were, by what principle can we draw metaphysical conclusions from epistemological premises? The fact that we can't empirically determine something does not imply that there is nothing to be determined. The world and our senses need not have been constructed in such a way that every physical fact can be settled by observation. Indeed, it is more of a miracle that we can reliably discover *anything* about the world by interacting with it than that we can't discover *everything* about it.

Of course, the inability to observationally determine some fact postulated by a physical theory is grounds for a certain sort of suspicion. If the supposed physical fact makes no difference to the observable behavior of the things, then we suspect that it can't really be doing any important work in the physical theory. Perhaps a new physical theory, purged of the inaccessible fact, could yield the same observable consequences at a lower ontological cost. Appeal to Occam's Razor might then militate in favor of the stripped-down theory over the original.

This is often easier said than done. Unobservable facts can be so entwined in the production of observable physical behavior, according to a theory, that the two cannot be separated. Leibniz's own attempt to purge absolute space from physics provides an example.

Newton identified the *observable* physical facts with the *sensible, relative* facts, such as the position and velocity of one observable bit of matter with respect to another. The absolute versions of these, the absolute motions, are not in the same way observable. And the relation between the observable facts and the unobservable is perfectly clear: the relative motion of two objects is determined by the difference of their absolute velocities. Even though the absolute velocity of particle 1 is unobservable, and the absolute velocity is particle 2 is unobservable, we are not free to assign to them whatever values we like: their relative motion is observable.

Leibniz's idea is to frame physics instead directly in terms of Newton's relative quantities: the relative locations and velocities of the material things. Leibniz would like to skim off the observable portions of Newton's ontology and leave the rest behind. But we already know Newton's argument against the feasibility of such a program: the bucket argument or, more directly, the argument of the rotating globes. If we accept that the globes, in an otherwise empty universe, could either be rotating or not and hence either display tension in the cord or not, the physical difference between these situations cannot be explained in terms of relative motions of bodies. All the material bodies in both cases are at relative rest.

Oddly, Clarke does not bring up Newton's argument until rather late in the correspondence (Fourth Reply, ⁋ 13), and Leibniz responds, apparently, by conceding Newton's fundamental claim:

> I grant there is a difference between an absolute true motion of a body, and a mere relative change of its situation with respect to another body. For when the immediate cause of the change is in the body, that body is truly in motion; and then the situation of other bodies, with respect to it, will be changed consequently, though the cause of that change be not in them. 'Tis true that, exactly speaking, there is not any one body, that is perfectly and entirely at rest; but we frame an abstract notion of rest, by considering the thing mathematically.[19]

It is entirely unclear how Leibniz thinks that this responds to Newton's argument. If the rotating bucket, or the rotating globes,

[19] Fifth Paper, ⁋ 53.

are in "absolute true motion" because of the (observable) presence of forces at work in them, then this absolute true motion cannot be a species of relative motion: there is none at all in the case of the globes. But if there is motion that is not the relative motion of bodies, and motion is change of location, then there are locations that are not relative to bodies, which just what Newton is arguing. So Leibniz never meets the challenge that Newton lays down, namely to account for the observable behavior of the bodies without invoking absolute space, time, and motion.

In the modern era, there have been several attempts to rise to Newton's challenge. The most famous is credited to Mach, although it falls short of being a full-fledged physical theory. It is clear that the Leibnizian relativist has an insurmountable task overcoming the globes example: if there is no relative motion of the globes to any other material body, no sort of "absolute motion" could be invoked by the relativist to account for different tensions in the cord. So Mach simply *rejects* the example. He says that we have no idea what would happen to such a pair of globes in an otherwise empty universe: the situation is too distant from reality for our actual experience of things to be any guide.

As for the actual experiment, the bucket experiment, Mach asserts that the shape of the surface of the water is accounted for by reference to the water's relative motion *with respect to the sphere of fixed stars*. It is clear that Mach cannot use any local relative motion: the rotation of the earth as a whole, for example, is evident in the bulge at the equator and the swirling motion of hurricanes. So if Mach is to find a relative motion that can account for the phenomena, he has to look far beyond our solar system. But what Mach never did provide was an actual theory: a dynamics couched in terms of relative motions that makes the sorts of predictions about observable phenomena that Newton's theory does. The most extensive investigation of how someone might try to frame physics in terms of relative quantities is the work of Julian Barbour and Bruno Bertotti.[20] This approach also has to reject the globes example and make it a necessary truth that the total mass of the universe is not rotating. Since Newton's theory allows

[20] Barbour and Bertotti (1982). See also Barbour (2000) and (2001).

for such rotation, the theory does not completely recover Newton's theory, even with respect to the physically possible relative motions of bodies.

Any evaluation of competitors to Newton's theory should begin with a clear statement of the drawbacks of Newtonian mechanics that the new theory means to avoid. For Mach, the drawback was philosophical: Mach thought that since the evidence on which physics rests is about the observable properties of observable bodies, physics itself should be formulated entirely in those terms. That led him to reject the atomic theory of matter as much as absolute space and time. But physics is evidently in the business of postulating unobservable entities in service of explaining observable behavior. The postulation is always risky, but, as the atomic hypothesis illustrates, the risk can sometimes pay off handsomely. Newton knew that absolute space and time are not, in themselves, observable, but he also explained how postulating them could help explain the observable facts. Why is this any worse than postulating atoms?

The unobservability of Newtonian absolute motion, though, takes a particularly worrisome form. We have already seen that the supposed unobservability of location in absolute space is hard to pin down, since it is not clear what sort of information an observation would be capable of conveying. If Clarke is right, the material universe *could have been* located elsewhere in absolute space—that is, located some other place than it is, keeping all the relative positions the same. But we do not need to make any observation to know that this did not *actually* happen: by hypothesis, the other placement of matter is counterfactual. However, there is a serious sort of unobservability inherent in Newton's theory, and we learn a lot by considering how these unobservable facts can be eliminated by changing the ontology of space and time.

Eliminating Unobservable Structure

Absolute Velocity and Galilean Relativity

In the preceding chapter, we focused on Clarke's first argument and Leibniz's jujitsu reversal of it. Clarke maintains that the facts about space must go beyond just the spatial relations between bodies, since bodies with the same spatial relations can be situated differently in absolute space; Leibniz responds that absolute space cannot exist, since (1) God could not choose between the different possibilities mentioned by Clarke, if they existed (PSR argument), and (2) the different possibilities, being qualitatively identical, can't really be distinct (PII argument). In each case, the two purported possibilities differ solely with respect to the *location* of the bodies in absolute E^3. But Clarke was not done. When Leibniz denied any real physical difference between the two situations, Clarke put forward another example where two distinct physical situations generate the same relative locations of material bodies:

> If space was nothing but the order of things coexisting; it would follow, that if God should remove in a straight line the whole material world entire, with any swiftness whatsoever; yet it would still always continue in the same place: and that nothing would receive any shock upon the most sudden stopping of that motion.[1]

Whereas the first of Clarke's examples concerns the *location* of matter in absolute space, the second example concerns instead the *absolute velocity* of matter in space. At one level, the idea is the same: if two situations can be physically different yet agree on the spatial relations of all bodies with respect to one another,

[1] Third Reply, ¶ 4.

then there must be something more to physics than just those relations. But in detail, the use of absolute velocity in the argument changes the situation drastically.

Leibniz responds to this argument in much the same way as to the first, mixing together the PSR and the PII even though the two principles apply in different ways:

> To say that God can cause the whole universe to move forward in a right [i.e., straight] line, or in any other line, without otherwise making any alteration in it; is another chimerical supposition. For, two states indiscernible from each other, are the same state; and consequently, 'tis a change without any change. Besides, there is neither rhyme nor reason in it. But God does nothing without reason; and 'tis impossible there should be any here. Besides, it would be *agendo nihil agere* [to act without doing anything], as I have just said now, because of the indiscernibility.[2]

The double lines of defense against Clarke are that (1) there really are not two distinct possibilities here, one with the material world at rest and the other with it moving uniformly in a straight line (PII argument) and (2) if, *per impossibile*, there were two distinct possibilities, God could have no grounds to choose between them (PSR argument). But on reflection, the PSR argument cannot get off the ground in the case of absolute velocity. Suppose that God wishes to create the material world in a heretofore empty absolute space. God could give the material world, as a whole, any absolute velocity in that space without affecting the relative positions and motions of bodies. Among all of these possible absolute velocities, one stands out as special: absolute rest. For if God should give the material world any nonzero absolute velocity, He would have to choose a *direction* in absolute space for that velocity to point. And the isotropy of E^3 implies that all of those different directions are qualitatively the same: there could be no reason to prefer one over another. The *only* absolute velocity that does not require a choice among directions is the zero velocity: absolute rest. So the PSR,

[2] Fourth Paper, ¶ 13.

far from prohibiting any choice of absolute velocity for the material world, demands exactly one choice.

In any case, as we have seen, the PSR argument relies on dubious theology, so we won't consider it further.

The PII argument, on the other hand, yields some interesting observations. First of all, both Leibniz and Clarke agree that uniform, straight-line motion of a material system through absolute space would be observationally undetectable. Clarke (and Newton) would deny this for the absolute *rotation* of a system, which is the point of the bucket argument. But a system in absolute uniform motion along a straight line will exhibit the same relative motions of its parts, and hence the same observable behavior, as it would at absolute rest. This claim, which is not intuitively obvious, was brilliantly defended by Galileo.

When arguing for the Copernican system, Galileo had to explain why the tremendously swift motion of the surface of the earth as the earth spins on its axis, and the even swifter motion of the earth as it orbits around the sun, would not be evident to casual observation. It helps to bear in mind just how fast these motions are, according to Copernicus. The circumference of the earth is over 24,000 miles, so at the equator the surface of the earth is moving at over 1,000 miles per hour due to the earth's rotation. And the orbit of the earth about the sun is about 585 million miles, so the earth travels over 1.6 million miles a day as it moves around the sun. These are almost incomprehensibly fast speeds. How could it possibly be that we don't *notice* that we are moving so fast?

Galileo pointed out that the absolute motion of a system, in certain circumstances, makes no difference to the observable behavior within the system. His own example used ships:

> For a final indication of the nullity of the experiments brought forth, this seems to me the place to show you a way to test them all very easily. Shut yourself up with some friend in the main cabin below decks on some large ship, and have with you there some flies, butterflies, and other small flying animals. Have a large bowl of water with some fish in it; hang up a bottle that empties drop by drop into

a wide vessel beneath it. With the ship standing still, observe carefully how the little animals fly with equal speed to all sides of the cabin. The fish swim indifferently in all directions; the drops fall into the vessel beneath; and in throwing something to your friend, you need throw it no more strongly in one direction than another, the distances being equal; jumping with your feet together, you pass equal spaces in every direction. When you have observed all these things carefully (though there is no doubt that when the ship is standing still everything must happen this way), have the ship proceed with any speed you like, so long as the motion is uniform and not fluctuating this way and that. You will discover not the least change in all the effects named, nor could you tell from any of them whether the ship was moving or standing still. In jumping you will pass on the floor the same spaces as before, nor will you make larger jumps toward the stern than toward the prow even though the ship is moving quite rapidly, despite the fact that during the time you are in the air the floor under you will be going in a direction opposite to your jump. In throwing something to your companion, you will need no more force to get it to him whether he is in the direction of the bow or the stern, with yourself situated opposite. The droplets will fall as before into the vessel beneath without dropping toward the stern, although while the drops are in the air the ship runs many spans. The fish in their water will swim toward the front of their bowl with no more effort than toward the back, and will go with equal ease to bait placed anywhere around the edges of the bowl. Finally the butterflies will continue their flights indifferently toward every side, nor will it ever happen that they are concentrated toward the stern, as if tired out from keeping up with the course of the ship, from which they will have been separated during long intervals by keeping themselves in the air.[3]

[3] Galileo (1967), pp. 186–187.

The claim that all experiments carried out in the two states of the ship would have the same observable outcome is called *Galilean Relativity*.

There is a tremendous amount of subtlety in the complete physical analysis of Galileo's experiment, as we will see when we get to the Theory of Relativity. But for the moment we need only grant that the experiments carried out in these two circumstances are in an obvious sense observationally indiscernible: no amount of experimentation below decks could reveal which state the ship is in. Galileo, like Newton and Clarke, takes it as obvious that the ship is in a *different state of motion* each time the experiments are carried out, but that the outcomes are not discernible because the *relative* positions and *relative* motions of all the bodies are the same in both cases. Relying on the postulate that the states of motion are different, Clarke uses the example to undermine Leibniz's claim that all motion is the relative motion of bodies.

Leibniz, of course, appeals to the PII argue that there are not really two different states at all. Or rather, Leibniz would concede that in the example Galileo gives, the ship in is different states of motion *with respect to the shore*, but Clarke demands something altogether different: that all the matter in the universe be put into uniform motion, so there is no difference in relative motion at all. This, Leibniz maintains, is nonsense.

Newton was aware of Galilean Relativity. Indeed, he proves this almost immediately in the *Principia* as Corollary V of the Laws of Motion, assuming that all forces are impulse forces due to collisions and that the magnitude of these impulses depends only on the relative speed of the bodies:

> *The motions of bodies included in a given space are the same among themselves, whether that space is at rest, or moves uniformly forwards in a right line without any circular motion.*
>
> For the differences of the motions tending towards the same parts, and the sums of those that tend towards contrary parts, are, at first (by supposition), in both cases the same; and it is from those sums and differences that the impulses and collisions do arise with which the bodies impinge on one another. Wherefore (by Law II), the effects

of those collisions will be equal in both cases; and therefore the mutual motions of the bodies among themselves in
the one case will remain equal to the motions of the bodies among themselves in the other. A clear proof of this we
have from the experiment of a ship; where all motions happen after the same manner, whether the ship is at rest, or is
carried uniformly forwards in a right line.[4]

We have already raised objections to the PII as a fundamental
principle of ontology or metaphysics but have conceded that the
postulation of unobservable physical facts should give us pause. In
Clarke's first argument, there did not appear to be any way to argue
that there is any physical question about the state of the world that
cannot be determined by observation. If we ask whether the material universe is oriented in absolute space as it is, with the eastern
parts in the east and the western parts in the west, or rather "quite
the contrary way," then the answer is clear: the universe is placed
where it is. Newton and Clarke admit that there is a distinct *counterfactual* possibility, in which all observable behavior would be
the same, but by its very formulation we know that the proposed
situation is counterfactual rather than actual.

But this second argument, which appeals to absolute motion
rather than position in absolute space, is quite different. For we
can now formulate physical questions about the material universe that, according to Newton, have determinate answers that
we cannot discover experimentally. For example, according to
Newton, right now you have a particular absolute velocity, with
a determinate speed and direction. You might be at absolute rest
(although because of the rotation of the earth, that could only
last an instant), or moving at one million miles per hour through
absolute space in the direction from the earth to Alpha Centauri.
Unlike the case of position, we do not know which of these descriptions of your absolute motion is correct, and, according to
Corollary V, no experiment can help.

So Newton's theory of absolute space and time, together with
his Laws of Motion and the postulate that all forces are impulse

[4] Newton (1934), vol. 1, pp. 20–21.

forces due to collisions, commit him to the existence of experimentally undeterminable physical facts. This should give us pause. Is there any way to somehow purge these unobservable absolute velocities from Newton's theory yet keep all of its explanatory power?

It is not obvious how this can be done. Galileo's observations, and Newton's corollary, do not entail anything like Leibniz's picture in which all motion is relative. Indeed, in both Galileo's and Newton's examples, the ships are postulated to be either "at rest" or "moving forwards in a right line without any circular motion," "not fluctuating this way and that." According to Newton, these are characterizations of the *absolute* motion of the ships. And it is only with respect to ships like this that Galilean Relativity holds. It is not the case, for example, that an experiment carried out in a lab will in general give the same result when carried out in a second lab in uniform motion with respect to the first. Indeed, it is not even true that an experiment will in general give the same result as another carried out in a lab *at rest* with respect to the first. For example, if I am experimenting on a spinning merry-go-round, seeing if drops of water will fall down from a bottle to a vessel beneath it, they will do so if the bottle is on the axis of rotation but not if it is out toward the edge of the merry-go-round. So the principle cannot be stated simply in terms of the relative motions of the labs: each lab, we would say, must itself be in *inertial* motion or constitute an *inertial frame of reference*.

To any student of physics, the phrase "inertial frame of reference" will be familiar. But what exactly does that phrase mean? Newton could give an account: an inertial frame of reference is a sensible body, relative to which the positions and motions of other bodies are referred, *that is at absolute rest or moving uniformly without rotation in a straight line through absolute space*. Such an account obviously does nothing to remove absolute space from the physics. So if we state Galilean Relativity as the claim that "all inertial frames are equal" and then define an inertial frame in this way, we are just as committed to absolute space and motion as Newton was.

At this point, we have a bit of a quandary. Newton's account of space and time commits him to facts about the absolute motion

53

of bodies that cannot be determined by any experimental means. But equally, whether a body is rotating or not, in some absolute sense, appears to have observable consequences. We would prefer to somehow eliminate the absolute motions but retain the absolute rotations, which seems to be a contradiction. As it turns out, the trick can be done. But we need to radically revise our approach to space and time.

Galilean Space-Time

Let's quickly review. Newton postulates that space has the structure of E^3 and that time has a one-dimensional metrical structure. And, most importantly, he postulates that the individual parts of absolute space persist through time, so there is a fact about whether a body at one time is in *the very same place* as it was at another time. If it is, the body is at absolute rest. If a body is not at absolute rest, there is a fact at every moment about how fast it is going and in what direction. Every body has a *trajectory through absolute space*: the points in the persisting E^3 that it occupies over an interval of time. That spatial trajectory has a shape: it could be a straight line, for example, or an ellipse. There is a further fact about how quickly the body passes through any given part of its spatial trajectory.

In terms of this ontology, Newton can define absolute rest and uniform absolute motion in a straight line, and hence can write his first Law of Motion. If we somehow want to retain Newton's dynamics but reject his account of space and time, we need some other structure in terms of which the First Law can be framed. Our route to this structure is through space-time.

We begin with a *representation* of Newton's ontology. For pictorial convenience, we will depict a Newtonian absolute space that is two-dimensional rather than three-dimensional. Suppose there are only two bodies, A and B, in this space. At any given moment of time, the bodies are located at some particular points in absolute space. As time goes on, the bodies can change their locations in absolute space. They might collide with each other and rebound, for example. Figure 5 represents the bodies at four

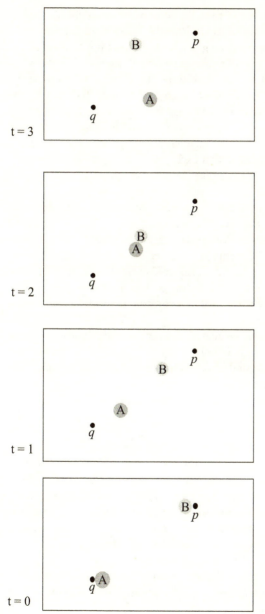

t = 3

t = 2

t = 1

t = 0

Fig. 5

55

different moments of time. In addition, two points of absolute space, labeled p and q, are indicated at all four moments.

The first thing we are going to do is stack up these representations of our two-dimensional space. Since there are infinitely many moments of absolute time, there ought to be infinitely many "leaves" in the stack, each representing the state of the world at that time. We will keep track particularly of our four. Furthermore, we will employ two obvious, but essential, conventions when stacking up our "pictures" of moments of time. First, we will stack them so that each point in absolute space forms a vertical line through time. The trajectories of the point p and q are indicated. Second, we stack them so the vertical distance between any pair of leaves is proportional to the elapsed absolute time. So if the elapsed time between $t = 0$, $t = 1$, $t = 2$, and $t = 3$ are all the same, the four leaves should be placed at equal intervals in space. The result is the three-dimensional *spatial* representation of our two-dimensional space as it persists through time. Since each of our colliding bodies also exists at each time, we can indicate their positions, forming the *world-lines* of these bodies (figure 6).

So far, we have not altered Newton's account of absolute space and time at all: we have only produced a purely spatial *representation* of the spatial and temporal structure that Newton posits. This sort of representation of both space and time is called a *space-time diagram*. It is essential to pay attention to the role that both absolute space and absolute time have played in producing the diagram: it is only because the points of absolute space persist through time that we could specify how to stack the time-slices, and only because absolute time has a metrical structure that we could specify the distances in space between the various slices. Objects at absolute rest are represented in this diagram by vertical lines, since they remain at the same point of absolute space. But much more importantly, *the trajectories of bodies in absolute uniform motion in a straight line through absolute space are represented by straight lines in the space-time diagram.*

Let's try one more diagram. In an otherwise completely empty space, there are two pairs of globes tethered by cords. One set is at absolute rest, and there is no tension in the cord. The other set is

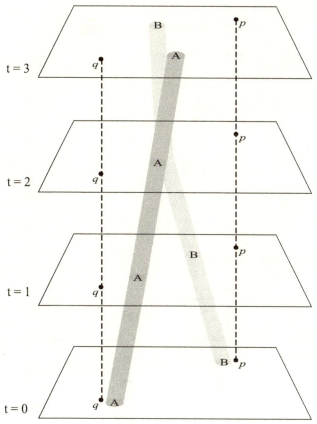

Fig. 6

in absolute rotation about its center of mass, and there is tension in the cord. Figure 7 is a space-time diagram of the situation.

Since the globes on the right in figure 7 are at absolute rest, their trajectories are represented by vertical straight lines in the diagram. The rotating globes, in contrast, are constantly changing their state of absolute motion. This is indicated on the diagram as a curvature of their world-lines. The absolute velocity of a globe at any moment corresponds to the tangent to the helix at that point. The absolute velocities of one of the globes at points r and s are

57

Fig. 7

represented by the two tangents shown in the diagram. The absolute speed of an object is indicated by the angle of this tangent: a vertical tangent corresponds to absolute rest, and the more tilted the tangent is from the vertical, the faster the absolute speed of the body. (N.B.: The absolute velocity is represented by the angle and orientation of the tangent; the "length" of the tangent has no significance.) The acceleration of the globes on the left is represented by the constant change in these tangents as time goes on. This continual alteration of the absolute motion of the globes is caused by the tension in the cord, in accordance with Newton's Second Law.

In this space-time diagram, *unaccelerated motions are represented by straight world-lines.* This holds both for bodies at absolute rest and bodies moving uniformly in a straight line. So in terms of this sort of diagram, we can rewrite Newton's First Law as follows:

> First Law (Space-Time Diagram Version): The trajectory of a body is represented by a straight line in a space-time diagram, except insofar as it is compelled to change its state by impressed forces.

Notice that in our new version, no distinction is made between bodies at absolute rest and bodies in uniform motion. Each of these is an *inertial motion*, that is, a motion that a body with no external forces on it can have. In our new version of the First Law, inertial motions are specified as exactly the motions represented by straight lines in a space-time diagram.

Newton's Second Law, in this language, would say that the effect of an impressed force is represented by a *curvature* or *bending* of the world-line in the diagram. The world-line will curve in the direction of the impressed force, and the amount of curvature will be proportional to the impressed force and inversely proportional to the mass of the body. The mass of a body, in particular the *inertial* mass of a body, is not mentioned in the First Law because it plays no role in the behavior of a force-free object. In essence, the inertial mass of an object is a measure of how much it resists having its world-line bent by forces: the bigger the mass, the less the trajectory bends. So schematically, instead of representing Newton's laws using

$$F = mA,$$

we can state the dynamics as

$$F = m\textbf{Bend},$$

where **Bend** is a measure of the curvature of the world-line. Putting a force on an object curves the world-line in the direction of the force, proportionally to the force and inversely proportionally to the inertial mass.[5]

What is essential about these reformulations of Newton's Laws is that all mention of the absolute velocity of any body has disappeared. Applying Newton's dynamics does not require that we make any distinction between absolute rest and uniform absolute motion: it only requires a distinction between *straight* and *curved* trajectories in space-time. As long as we can identify the straight lines in our space-time diagram, and quantify the curvature of the curved lines, we can determine what Newton's Laws predict in any given case. *We don't need to know the trajectories of individual points of absolute space, as indicated in figure 6.*

And what we don't need to know, we don't need. That is, although Newton postulates that the individual points of absolute space persist through time, we have managed to reformulate his Laws without any reference to that persistence or to the absolute velocity of anything. We are free to reject Newton's account of space and time in favor of a new one: Galilean space-time. Of course, we cannot reject the persistence of points of space through time without paying some price elsewhere: we must do more than just eliminate that bit of metaphysics. And in order to add in the geometrical structure we need, our entire account of the ontology of space and time has to change.

Newton postulates two distinct sorts of things: absolute space, which has infinitely many points, and absolute time, which has infinitely many moments or instants. Because the points of absolute space persist through time, the same set of points constitutes space at all times. Our new ontology rejects this account in favor of *space-time points* or *events*. An event is essentially a place-in-space-at-a-time: the explosion of a particular firecracker takes place at a space-time point, which occurs only once and, ideally, has no spatial extension and takes up no time. Learning to

---

[5] This law obviously requires that one be able to quantify the amount of curvature of a line and determine the direction of the curvature. This can be done, although explaining the mathematical techniques would needlessly detain us.

think about space and time in terms of events requires some practice, and our space-time diagrams can be of great assistance here. An event corresponds to a single point in a space-time diagram, whether anything of note is happening there or not.

Once we accept events as the basic elements of our spatiotemporal ontology, our job is to specify the physical structure of these events: the geometry of space-time. Galilean space-time postulates several different sorts of structure.

First, there is an absolute, objective lapse of time between any pair of events. In this sense, Galilean space-time adopts Newton's absolute time without any change. There is a metric for the elapsed times, so it is well defined, for example, whether the time between one pair of events is the same as, greater, or less than the elapsed time between any other pair.

Given this absolute temporal structure, Galilean space-time can be partitioned into sets of events that all happen at the same time, that is, events between which the lapse of time is zero. Each such complete set of events forms a *simultaneity slice* in the space-time. Figures 6 and 7 indicate these simultaneity slices in an obvious way. The elapsed time between any event on the slice marked $t = 0$ and any event on the slice marked $t = 1$ in figure 2 is the same. We call this partitioning of the space-time into simultaneity slices a *foliation* of the space-time. So the structure of time foliates Galilean space-time into simultaneity slices in a completely objective way. For any pair of events, there is a fact about which, if either, occurs first and how much time elapses between them.

That takes care of the temporal structure. What about space? Galilean space-time postulates a particular spatial geometry *on each simultaneity slice*. In particular, it postulates that the events on each slice have the spatial structure of E^3. The geometry of each slice has to be specified independently of every other because the points of space do not persist through time. So indicating that the events on the slice $t = 0$ have the structure of E^3 tells us nothing, in itself, about the geometry of the events on $t = 1$. The second piece of geometrical structure of our space-time, then, is the spatial geometry E^3 on each individual slice.

But just postulating absolute time and E^3 on each simultaneity slice does not give us enough structure to state Newton's Laws of

Motion. As we have seen, Newton's Laws require an additional bit of *trans-temporal* geometry: geometrical structure among events at *different* times. For Newton himself, the trans-temporal structure was secured by the persistence of points of absolute space through time, but we are trying to eliminate that postulate. To take its place, though, we need a topology and an affine structure in our new space-time. The topological structure determines which sets of events constitute unbroken, continuous trajectories through time. And the affine structure determines which of these continuous space-time trajectories are straight. (By a "continuous space-time trajectory," we mean a continuous set of events that contains at most one event at each time.)

The idea that space-time (i.e., the set of all events) has a topology is so natural and intuitive that we are hardly aware of it as a postulate. When we draw space-time diagrams, we have to draw them in a space (the page) that has a topology, an affine structure, and a metric. But now that we are postulating a structure for space-time itself, we are not entitled to assume that it has the same geometry as the diagram. In fact, we will soon see that the space-time diagrams have a *different* geometrical structure than the space-time they are meant to represent, so we have to learn to systematically *ignore* some aspect of the picture when interpreting the diagrams. The trajectories of bodies through space-time, their world-lines, can only be characterized as continuous or discontinuous if the space-time has a topology, so we postulate it has one. Indeed, the topology of the space-time corresponds to the topology of the space-time *diagram*, so continuous trajectories will be represented by continuous curves in the picture.

But in addition to the topology, Galilean space-time contains an affine structure, so that world-lines can be characterized as either straight or curved. This affine structure is called the *inertial structure* of the space-time, for obvious reasons. A body subject to no external forces is in *inertial motion* and so has an *inertial trajectory* through space-time. In Galilean space-time, the inertial trajectories are just the straight trajectories. Given this space-time structure, then, it is possible to formulate Newton's Laws without postulating the persistence of points through time and, hence, without postulating any absolute velocities. Our puzzle of how there can be absolute *acceleration* (e.g., the rotation of the globes in figure 7) without any absolute *velocities* is solved: absolute acceleration is the bending of a world-line in space-time.

I mentioned above that the geometrical structure of a space-time *diagram* is not the same as the geometrical structure of the space-time being represented. What we are doing when we draw a diagram is using one kind of geometrical object to represent another. Since the geometries of the two objects are not the same, we must pay close attention to which aspects of the diagram correspond to real physical facts and which are merely conventions. In the case of Newton's own account of space and time, we have seen that the *angle* of a trajectory on the diagram has a physical significance: it represents the *absolute velocity* of a body, with objects at absolute rest occupying vertical trajectories. But in Galilean space-time, there are no absolute velocities. It is a matter of arbitrary choice which straight trajectories are depicted as vertical and which as "tilted" in the diagram. So for example, the real physical situation in Galilean space-time depicted in figure 7 could just as accurately have been depicted as in figure 8.

Figure 8 is related to figure 7 by an affine transformation, like the one depicted in figure 1. Under this transformation of the *diagram*, all of the Galilean structure remains unchanged: the elapsed times between events are the same, the simultaneity slices are the same, the Euclidean structure of each individual slice is the same, and the set of straight space-time trajectories is the same. The "state of motion" of a body at a time is still represented by the tangent to its trajectory at that time, so the state of motion of the globe at *r* still differs from its state of motion at *s*. This difference is attributable, in exactly the same way, to the tension in the cord. So if we postulate that physical space-time has a Galilean structure, figures 7 and 8 can be used to depict *exactly the same physical situation*.

Using our conventions for depicting full Newtonian space-time, the structure that Newton posits, figures 7 and 8 would have to depict *different* physical situations: in figure 7 the globes on the right are at absolute rest and in figure 8 they are moving at a constant, nonzero absolute velocity. The *accelerations* of the bodies in the two situations are the same, so the *forces* at work are the same

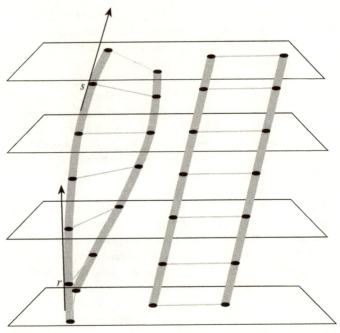

Fig. 8

(same tension in the cord), but for Newton these are physically different states of affairs. In Galilean space-time, they are not.

The Galilean space-time structure is presupposed in all standard accounts of Newtonian mechanics, even though this presupposition is usually tacit and unremarked. For example, in the account of Newtonian mechanics given by Goldstein above, the presupposition occurs almost immediately, although we would never notice. Goldstein begins: "Let **r** be the radius vector of a particle from some given origin. . . ." But *what* "given origin"? If we *randomly* pick a new origin at each moment of time, then the radius vector will change randomly, no matter how the particle behaves. If what is meant is to always use *the same* origin at all times, then the account presupposes Newtonian absolute space persisting through time. But what is really supposed is that the trajectory of the origin of the coordinate system should be an *inertial* trajectory: the spatial origin of the coordinate system should trace out a straight line through space-time.

Indeed, we now have the resources to say, in general, what an *inertial coordinate system* in Newtonian mechanics is. In Galilean space-time, with the space being E^3, a coordinate system will assign four numbers to each event: three "spatial coordinates" and one "time coordinate." The differences in values of the time coordinate should be proportional to the elapsed time between events. So all the events that take place on a single simultaneity slice should be assigned the same time coordinate. And *on each time slice*, the three spatial coordinates should form a Cartesian coordinate system on E^3. Spatial distance, and other purely spatial geometry, only obtains among events that occur at the same time. These requirements on a coordinate system are fairly obvious. But in order to be an *inertial* coordinate system, something more is required: the coordinate curves associated with the time variable must be inertial trajectories. Suppose we are using the four coordinates (t,x,y,z). Fix the values of the three spatial coordinates, and let the t coordinate vary. For example, consider all the events whose coordinates are $(t,2,3,5)$, as t varies. This set of points will form a continuous line in the space-time (assuming the coordinate functions are continuous). In an inertial coordinate system, all of these coordinate curves will be straight.

That is why "attaching a coordinate system" to a rotating body yields non-inertial coordinates. If, for example, we used one of the globes on the left in figure 7 as the origin of a coordinate system, the resulting coordinates would not be inertial. The critical point to notice is that *the distinction between inertial and non-inertial coordinate systems is parasitic on a prior distinction between inertial and non-inertial trajectories in space-time itself.* It is only because the physical space-time has the right kind of affine structure that a meaningful distinction among coordinate systems is possible.

Galilean space-time provides the ideal arena for Newtonian mechanics. There is enough space-time structure to express Newton's laws of motion, enough to distinguish rotating from non-rotating systems and so explain the phenomena in the bucket experiment, enough to define inertial and non-inertial coordinate systems. But there is *not* enough structure to define absolute rest or absolute velocity in general. So the embarrassing question that Newton could not empirically answer—what is your present absolute velocity?—cannot be posed in Galilean space-time. If Newtonian mechanics had turned out to be adequate to explain all observable phenomena, physicists would almost certainly have settled on Galilean space-time as the correct account of spatio-temporal structure. We could, of course, insist on the full Newtonian picture with points of space persisting through time, but that extra structure buys no more explanatory power. And we could, with Leibniz and Mach, somehow try to eliminate space-time as an entity altogether, but it is unclear what either the motivation or the prospects of such a project are. Space-time structure is not directly observable, but it nonetheless plays an essential role in the formulation of physical theory. There is no more call to try to eliminate space-time structure from physics than there is to eliminate the postulation of atoms, just because they cannot be directly seen.

But, of course, all was not well with Newton's theory. Various phenomena, especially phenomena involving light and electromagnetism, resisted all attempts to account for them in Newtonian terms. As result, Newtonian mechanics was overthrown in favor of the Special Theory of Relativity, and both Newton's absolute space and time and Galilean space-time were overthrown with it.

Special Relativity

SPECIAL RELATIVITY AND MINKOWSKI SPACE-TIME

Special Relativity is a very simple theory that is commonly presented in a complex and confusing way. Some of the reasons for this are historical. Einstein himself presented the theory as the consequence of two principles: (1) the equivalence of all inertial frames and (2) the constancy of the speed of light. From these two principles, after some fashion, we derive the *Lorentz transformations*, which are a set of equations relating one set of coordinates to another. But already we can see that this approach to understanding the theory has run seriously off the rails.

First, the notion of an inertial system, or an inertial set of coordinates, or an inertial frame of reference, is derivative rather than fundamental. Each of these concepts can only be defined by reference to some objective geometrical structure of space-time itself, in order to make sense of the qualifier "inertial." So we ought to begin with the intrinsic geometry, not with coordinate systems or reference frames. Second, we have just gone to great lengths to eliminate absolute velocities from Newtonian mechanics. We were very pleased that in Galilean space-time, nothing has an absolute speed. So it ought to strike us as going backward to found Special Relativity on claims about the speed of anything. Basing a presentation of Relativity on talk of speeds unavoidably suggests that we are dealing again with Newtonian absolute space and time and Newtonian absolute motion. Popular presentations of Relativity trade on this tacit suggestion all the time. They say things like "In Relativity, when an object goes fast, time slows down" or "In Relativity, as an object approaches the speed of light it shrinks" or "We don't notice relativistic effects in everyday life because we don't travel anywhere near the speed of light." Each of these claims implies that a body, at any given moment, *has* a speed that can be closer or further from the speed of light. But in Relativity, just as in Galilean space-time, there simply are no such speeds. There is no physical fact about how fast the earth is moving right now; it is no more correct to say that it is not traveling near the speed of light than that it is traveling at 99 percent of the speed of light. To understand Relativity, we have to expunge all ideas of things having speeds, including light.

To begin with, we will focus on just a single physical phenomenon that can be checked experimentally. The phenomenon is that *the trajectory of light in a vacuum is independent of the physical state of its source*. In particular, suppose there are two flashbulbs in very rapid relative motion, moving past each other. As the flashbulbs pass each other, they both go off. It is an empirically verifiable fact that the light from the two flashbulbs will reach an observer anywhere in the universe exactly together. So the trajectories of the light, the paths the light follows, do not depend on what the source was doing when the light was emitted.

There are other aspects of the light from the two flashbulbs that can be different. For example, observers will typically record the light as having a different *color* or *wavelength* or *frequency*, depending on which bulb it came from. But nonetheless the two rays of light will arrive at the observer together. This is clearly a necessary condition if we are to maintain that "the speed of light is constant": if one light ray could outrun or overtake another, then their speeds could not be considered to be the same. But the phenomenon we just described does not mention the speed of anything or depend upon being able to observe or calculate the speed of anything. We will use this physical fact as one touchstone.

If we accept that in a vacuum there is no physical structure except for the structure of space-time itself, then the behavior of light in a vacuum implies that *the geometry of space-time alone determines the trajectory of light rays*. That is, given any point in space-time *p*, the structure of space-time ought to fix where light emitted from that *p* (in any possible direction) will go. This set of points, the places where light emitted from *p* (in a vacuum) might end up, is called the *future light-cone of p*. And similarly, any light that manages to reach p must have come along one of a particular set of trajectories that form the *past light-cone of p*. So our simple observation about the behavior of light in a vacuum implies that the geometry of space-time, whatever it is, ought to associate with each event a past and a future light-cone.

At this point, instead of trying to somehow derive Special Relativity from some phenomena or some general principles, we will simply state, in as clear a way as possible, what sort of space-time structure Special Relativity postulates to exist. Given that postulate, we will then consider some physical claims that can be framed in terms of that geometry. As it turns out, the claim "The speed of light is constant" is physically very complicated, and we can only make clear sense of it once a lot of other work has been done.

In order to specify the geometrical structure of Special Relativistic space-time, we will follow the same strategy used above for E^3. We saw that the geometry of E^3 can be described a bit indirectly by reference to Cartesian coordinate systems: a space is E^3 if and only if it *admits of* coordinates that relate to the geometrical structure in a particular way. The space-time of Special Relativity is called *Minkowski space-time*, and its geometry can also be specified indirectly by reference to some special coordinate systems, called *Lorentz coordinates*. In fact, the way that Lorentz coordinates relate to the geometry of Minkowski space-time is almost the same as the way Cartesian coordinates relate to Euclidean space. There is just one seemingly small difference between the two cases. But we should always keep in mind Einstein's warning: coordinates need not have any direct *physical* significance. In particular, one of the coordinates will be called the t-coordinate, which suggests that it has something to do with *time* or with *clocks*. But we are not presupposing any such connection: right now, the coordinates are nothing but numbers that get assigned to events.

Minkowski space-time is four-dimensional, that is, we need four coordinates to cover it with continuous coordinate functions. The coordinates are traditionally called t, x, y, and z. Since the coordinate functions are all continuous, this already determines the topology of Minkowski space-time: a point moves around continuously in Minkowski space-time if and only if its Lorentz coordinates all change continuously as it moves. Since this is exactly the same as Cartesian coordinates on E^4, Minkowski space-time and four-dimensional Euclidean space are topologically identical.[1]

The affine structure of Minkowski space-time also has exactly the same relation to Lorentz coordinates as the affine structure of Euclidean space has to Cartesian coordinates. That is, a set of events in Minkowski space-time forms a complete straight line if and only if, for some choice of numbers A, B, C, D, E, F, G, and H, the coordinates of the points in the set have the values

$$(As + B, Cs + D, Es + F, Gs + H),$$

where s ranges over the real numbers. As in the case of Euclidean space, at least one of A, C, E, and G must be nonzero.

In sum, both the topological and straight-line structure of Minkowski space are the same as E^4. This means that Euclidean space-time *diagrams* can represent these aspects of Minkowski space-time in a particularly simple way. Continuous lines on the diagram can correspond to continuous lines in the space-time, and straight lines on the diagram can correspond to straight lines in the space-time. Of course, we can't produce actual four-dimensional diagrams, but we can depict three-dimensional Euclidean space in the usual way. So if we suppress one of the dimensions of Minkowski space-time, the remainder can be depicted straightforwardly and accurately. That leaves only the metrical structure of Minkowski space-time to consider.

Given Cartesian coordinates on E^3, the metrical structure is represented, in terms of the coordinates, by the function

$$D(p,q) = \sqrt{\left(X(p) - X(q)\right)^2 + \left(Y(p) - Y(q)\right)^2 + \left(Z(p) - Z(q)\right)^2},$$

where $X(p)$, $Y(p)$, and $Z(p)$ are the three coordinate functions of the Cartesian coordinates. In Minkowski space-time, the analogous geometrical structure is called the *Invariant Relativistic Interval*, and it can be represented, in terms of Lorentz coordinates as

---

[1] This is the accepted wisdom on the matter. I myself dispute this; see chap. 1, n. 2, above.

$$I(p,q) =$$
$$\sqrt{\left(T(p) - T(q)\right)^2 - \left(X(p) - X(q)\right)^2 - \left(Y(p) - Y(q)\right)^2 - \left(Z(p) - Z(q)\right)^2}$$

(Equation 2).

We have just completed our account of Minkowski space-time. All the rest is analysis of this geometry.

The heart of Special Relativity is contained in equation 2, and we will spend a lot of time unpacking its significance. But first a few caveats, especially for readers who have studied physics.

In many physics texts, the Interval is defined as the square of the quantity given above. But the square root is more appropriate, because the ratios of magnitudes in Minkowski space-time are proportional to the ratios of the quantity defined in equation 2. The choice of the square root has what may seem to be very odd consequences: for example, the Interval as defined is sometimes an *imaginary* number. But since all we really care about are the ratios among these numbers, that presents no difficulty. Also, in some texts a different convention is used, and the square of the difference in the t-coordinate is subtracted from the sum of the squares of the differences of the other coordinates. Changing equation 2 in this way would result in changing all real values of the Interval to imaginary values, and vice versa. But again, since all that matters are the ratios among these numbers, this is no real difference.

Finally, in many physics texts, the definition is given for a differential version of the Interval. That is, if the points p and q are sufficiently close to one another, their coordinate values will be close to one another. So, for example, $T(p) - T(q)$ will be a very small number, which we can designate dT. In differential form, equation 2 becomes

$$dI = \sqrt{dT^2 - dX^2 - dY^2 - dZ^2}.$$

Mathematically, the differential form is much more powerful and must be used in the General Theory of Relativity. But for our exposition, we will use equation 2, because we can calculate with it easily.

The quantity $I(p,q)$ defined in equation 2 is sometimes called the *Minkowski metric*, but this is can cause some confusion. The function does not meet the definitional requirements for a metric function: it is not positive-definite and does not satisfy the triangle inequality. This should serve as a warning that the geometry of Minkowski space-time is fundamentally quite unlike the geometry of Euclidean space, despite the similarity in topological and affine structure. So while the topological and straight-line structure in a space-time diagram represent just what they seem to, *distances between points* in a space-time diagram do not directly correspond to the *Intervals between events* that are represented. This must always be kept in mind when interpreting a relativistic space-time diagram.

It is a mathematical fact that if a space admits of one set of Lorentz coordinates, it admits of infinitely many such coordinate systems, just as a Euclidean space admits of infinitely many Cartesian coordinate systems. These coordinate systems vary with respect to location of the origin, orientation of the coordinate axes, and scale. Given any pair of Lorentz coordinate systems, there will be a set of equations that transform the coordinates of an event in one system to its coordinates in another system. The set of transformations that agree on the origin and scale of the coordinates are called *Lorentz transformations*; a wider set that only have to agree on scale are called *Poincaré transformations*. The choice of coordinate axes in a Lorentz coordinate system is not quite as free as the choice of axes for Cartesian coordinates in Euclidean space: any straight line in Euclidean space can serve as a coordinate axis of a Cartesian system, but certain straight lines in Minkowski space-time cannot serve as coordinate curves in a Lorentz frame. Minkowski space-time is not isotropic in the same way as Euclidean space. Different directions in Minkowski space-time can have a different geometrical flavor. This allows Minkowski space-time to explain the behavior of light.

The function that defines the Interval between events is not positive-definite, and the function takes the value zero for some pairs of distinct points. For this reason alone, the Interval cannot accurately be considered a *distance*: any pair of distinct points should have some nonzero distance between them. To get a sense of the structure of these zero-valued Intervals, we will draw a space-time diagram of Minkowski space-time with one dimension 72 suppressed. Since the topology and affine structure of Minkowski space-time are the same as the topology and affine structure of Euclidean space, our diagram will look like E^3. Furthermore, we will use Cartesian coordinates on this E^3 to represent the Lorentz coordinates on Minkowski space-time. So instead of labeling the coordinates x, y, and z, we will label them x, y, and t. By convention, the t-coordinate runs vertically on the page, and the x and y coordinates are at right angles. Consider the origin o of the coordinate system, the event with coordinates $(0,0,0,0)$. In order for some other event p to have a zero Interval from the origin, we must have

$$I(p,0) =$$
$$\sqrt{(T(p) - T(o))^2 - (X(p) - X(o))^2 - (Y(p) - Y(o))^2 - (Z(p) - Z(o))^2} =$$
$$\sqrt{(T(p) - 0)^2 - (X(p) - 0)^2 - (Y(p) - 0)^2 - (Z(p) - 0)^2} =$$
$$\sqrt{(T(p))^2 - (X(p))^2 - (Y(p))^2 - (Z(p))^2} = 0.$$

That is, the set of events at Interval zero from the origin are the points p whose coordinates satisfy the equation

$$t_p^2 - x_p^2 - y_p^2 - z_p^2 = 0.$$

If we draw this set of events on a space-time diagram (suppressing the z-dimension), they form a double cone whose cone-point is the origin. The intrinsic geometry of Minkowski space-time associates such a double cone with every event. Two such cones are depicted in figure 9.

In the Theory of Relativity, the behavior of light in a vacuum is associated with these cones, hence they are called *light-cones*. In particular, Relativity (or, more exactly, the relativistic theory of electromagnetism) entails the

> Law of Light: The trajectory of a light ray emitted from an event (in a vacuum) is a straight line on the future light-cone of that event.

If a flashbulb goes off at o (in a vacuum) the light, spreading out in all directions, will occupy the surface of its entire future light-cone. Intuitively, in three-dimensional space, a flashbulb produces an expanding sphere of light. With one spatial dimension suppressed, this becomes an expanding circle of light, which appears as a cone in our diagram.

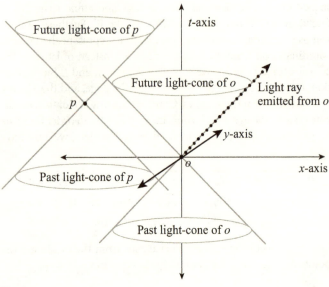

Fig. 9

The Law of Light can only be formulated in a space-time that associates a light-cone with each event. Note that the Law of Light mentions nothing about the source of the light save that the source emits at a particular event. So the Law of Light implies the phenomenon cited above: two light rays emitted from the same point in a vacuum will arrive together at a distant observer. We have accounted for this phenomenon without defining, or even mentioning, the "speed of light."

The light-cone of *o* partitions the remainder of the universe into five topologically connected sets: the events within the future light-cone, the events on the future light-cone, the events within the past light-cone, the events on the past light cone, and the events outside the light-cone. Events within the past or future light-cone are called *time-like separated* from *o*, those on the future or past light-cone are *light-like separated* from *o*, and those outside the light cone are *space-like separated* from *o*. Given our conventions, in any Lorentz coordinate system, for all the time-like separated pairs of events, the Interval $I(p,q)$ is positive and real; for light-like separated events, $I(p,q)$ is zero; and for space-like separated events, $I(p,q)$ is imaginary. All Lorentz frames agree on the ratios among these intervals.

The claim that *nothing can go faster than light* can be expressed in terms of this geometry as follows:

> Limiting Role of the Light-cone: The trajectory of any physical entity that goes through an event never goes outside the light-cone of that event.

Once again, we did not need to define the speed of anything to state the physical principle.

Since Minkowski space-time has an affine structure, Newton's First Law can take the same form as it did in Galilean space-time:

> Relativistic Law of Inertia: The trajectory of any physical entity subject to no external influences is a straight line in Minkowski space-time.

Given that a light ray in a vacuum is subject to no external influences, it follows that the trajectory of light in a vacuum is straight. But the Relativistic Law of Inertia also governs massive bodies whose trajectories do not lie on the light-cone. Indeed, it is a postulate of Relativity that the trajectory of any massive body is everywhere time-like, and so always lies inside the light-cone of every event it passes through.

The Relativistic Law of Inertia, together with the fact that the affine structure of Minkowski space-time is identical to that of E^4, implies that in a space-time diagram the image of the trajectory of a force-free object will be a straight line. We will use this fact extensively when solving Relativistic problems.

Given the geometrical structure of Minkowski space-time, the Relativistic Law of Inertia already allows us to make some predictions. For example, if the trajectories of two free physical entities (massive bodies or light rays) intersect, they intersect only once. This follows from the fact that in Minkowski space-time, no pair of straight lines intersects more than once. This may seem to be a rather trivial prediction, but we will eventually see that it is no longer true in General Relativity.

In order to get more impressive physical predictions from the theory, we need more connections between the geometry of Minkowski space-time and the observable behavior of material objects. The most important such connection is a *hypothesis*, whose ultimate status will be scrutinized later:

> Clock Hypothesis: The amount of time that an accurate clock shows to have elapsed between two events is proportional to the Interval along the clock's trajectory between those events, or, in short, clocks measure the Interval along their trajectories.[2]

The Clock Hypothesis holds the key to understanding all of Relativity.

The very notion of a "clock" in Relativity ought to be a bit of a puzzle. After all, when constructing Minkowski space-time, we nowhere postulated anything like Newtonian absolute time. Unlike Galilean space-time, Minkowski space-time is not foliated into simultaneity slices; indeed, the very notion of "simultaneous events" has no content at all. The light-cone structure somehow replaces the foliation. Since there is no absolute time in Relativity, clocks certainly cannot measure it. But clocks must be measuring *something*: two accurate clocks, placed side by side, will tick in unison, so there must be some geometrical structure in space-time that both of the clocks are reflecting. According to the Clock Hypothesis, that structure is the Interval. Accurate clocks, in Relativity, are like odometers on cars, measuring the length of their trajectory through space-time. If we grant this, then clocks that are side by side must tick together, since their trajectories through space-time will be the same length. But clocks that wander off on different trajectories can record quite different elapsed times between the same pair of events, just as cars that take different routes between the same locations can show different elapsed miles on their odometers.

---

[2] It is here that our definition of the Interval using the square root redeems itself: using the more standard convention, clocks measure the square root of the Interval.

THE TWINS PARADOX

Let's use the Clock Hypothesis to explain the iconic phenomenon of Relativity: the so-called Twins Paradox. Qualitatively, the situation is this: two twins, with identically constructed clocks, begin in a situation where they are side by side in rocket ships and subject to no forces. Twin A briefly turns on his engines, then turns them off. The twins drift apart. After a while, twin A again fires his engines, but in the opposite direction. He eventually drifts back to twin B, who has never fired his engines. Twin A fires his engines a third time, coming back to relative rest with respect to twin B. When the twins compare their clocks, they find that the twin B's clock has run off more time than twin A's. Furthermore, twin B appears to be biologically older than twin A. How does Special Relativity predict this?

In order to a get quantitative account of the phenomenon, we have to specify the exact trajectories of the two twins, and for this purpose we will use a conveniently chosen Lorentz coordinate system. The convenience of the system plays no role in the explanation: any other Lorentz frame would give the same results, but the calculations would be more complicated. In this convenient coordinate system, twin B's trajectory is always along the t-axis: his x-, y-, and z- coordinates are always 0. Twin A fires his rockets for the first time at the origin o. He then moves inertially until he reaches the event p whose coordinates are (5,4,0,0). He fires the rockets again, and moves inertially until meeting his brother at event q, whose coordinates are (10,0,0,0). When they get back together, twin B's clock indicates that 100 days have elapsed. How much time has elapsed according to twin A's clocks?

To solve this problem, we will use a space-time diagram. The inertial motions of the twins will be represented by straight lines on the diagram, in accordance with the Relativistic Law of Inertia. Since the y- and z-coordinate values of both twins are always zero, we will ignore those dimensions and use a two-dimensional diagram, figure 10.

In order to determine how much time will have elapsed according to twin A's clock, we need to know the ratio of the length of twin B's trajectory compared with twin A's trajectory. To do

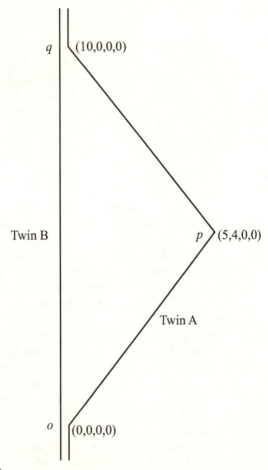

Fig. 10

this, we will make use of the quantity I as calculated in these Lorentz coordinates. The length of the straight line $\overline{oq}$ is

$$\sqrt{(T(q) - T(o))^2 - (X(q) - X(o))^2 - (Y(q) - Y(o))^2 - (Z(q) - Z(o))^2} =$$
$$\sqrt{(10 - 0)^2 - (0 - 0)^2 - (0 - 0)^2 - (0 - 0)^2} =$$
$$\sqrt{100} = 10.$$

The length of $\overline{op}$ is

$$\sqrt{(T(p)-T(o))^2 - (X(p)-X(o))^2 - (Y(p)-Y(o))^2 - (Z(p)-Z(o))^2} =$$
$$\sqrt{(5-0)^2 - (4-0)^2 - (0-0)^2 - (0-0)^2} =$$
$$\sqrt{25-16} = \sqrt{9} = 3.$$

The length of $\overline{pq}$ is

$$\sqrt{(10-5)^2 - (0-4)^2 - (0-0)^2 - (0-0)^2} =$$
$$\sqrt{25-16} = \sqrt{9} = 3.$$

So the total length of A's trajectory, in these units, is $3 + 3 = 6$, while the total length of B's trajectory is 10. If B's clock shows 100 days of elapsed time, A's shows 60. When the twins get back together, A will be 40 days younger than B.

One reason the twins' situation may seem "paradoxical" is that anyone looking at figure 10 would naturally conclude that A's path from o to q is *longer* than B's path. And indeed, for the representations *on the diagram* this is true. But the diagram is drawn in Euclidean space, so we have to be very cautious when interpreting it. In Minkowski space-time, the Triangle Inequality does not hold for the Interval.[3] Rather, the Interval obeys a sort of anti–Triangle Inequality: the sum of the lengths of A's two sides of the triangle in figure 10 will always be *less* than the length of the remaining side.

The Twins "Paradox" has inspired more confusion about Relativity than any other effect. The explanation of the phenomenon, in terms of the intrinsic geometry of Minkowski space-time and the Clock Hypothesis is exquisitely simple: clocks measure the Interval along their world-lines, and B's world-line between o and q is longer than A's. Period. There is nothing more to say.

Let's clear up some of the confusions.

Confusion 1: In Relativity, all motion is the relative motion of bodies. But the relative motion of A with respect to B is exactly the same as the relative motion of B with respect to A. So their physical situations are exactly the same.

---

[3] This is another reason the Minkowski "metric" is not a metric.

Therefore, when they get back together, by symmetry, their clocks must show the same elapsed time.

Every claim made in Confusion 1 is incorrect. First, it is not true in any interesting sense that "all motion is relative in Relativity." For example, the explanation of Newton's rotating globes in Special Relativity is essentially *identical* to the explanation given by Newton himself or the explanation given in Galilean space-time: there is an objective geometrical structure to space-time that defines inertial trajectories for all bodies, irrespective of the existence of any other bodies. The tension in the cord in one case indicates that the globes are not following inertial trajectories: their world-lines are curved. The space-time diagram for the rotating globes in Special Relativity would look just like figure 7, except that in place of the simultaneity slices we would have light-cone structure. Relativity is not a "relationist" theory of space-time of the sort Leibniz sought.

> Confusion 2: As a "solution" to Confusion 1, many texts suggest that the Twins Paradox is resolved by *acceleration*: twin A's situation is not really symmetrical with twin B's because twin A has had to use his rockets. The rockets (in accordance with the Relativistic version of Newton's Second Law) produce a force that twin A can feel and that bends his world-line. Twin B does not use his rockets and feels no force. So twin B knows it is he who is "really at rest" while his twin, feeling the force, knows he "really travelling."

Confusion 2 is so prevalent that it would impossible to cite all examples, but two can give the flavor. Wolfgang Rindler, in his text *Essential Relativity*, writes

> Now the paradox is this: cannot *A* . . . say with equal right that it was *he* who remained where he was, while *B* went on a round-trip, and that, consequently, *B* should be the younger when they meet? The answer, of course, is no, and this resolves the paradox: *B* has remained at rest in a single inertial frame, while *A*, in the simplest case of a uniform to-and-fro motion—say from earth to a nearby star and back—must at least be accelerated briefly out of *B*'s frame into another, decelerated again briefly to turn around, and

finally decelerated to stop at *B*. These accelerations (positive and negative) are *felt* by *A*, who can therefore be under no illusion that it was he who remained at rest.[4]

The great Richard Feynman, in his *Lectures on Physics*, tells a similar tale:

> This is called a "paradox" only by people who believe that the principle of relativity means that that *all motion* is relative; they say "Heh, heh, heh, from the point of view of Paul can't we say that *Peter* was moving and should therefore appear to age more slowly? By symmetry, the only possible result is that both should be the same age when they meet." But in order for them to come back together and make the comparison, Paul must either stop at the end of the trip and make a comparison of clocks, or, more simply, he has to come back, and the one who comes back must be the man who was moving, and he knows this, because he had to turn around. When he turned around, all kinds of unusual things happened in his space-ship—the rockets went off, things jammed up against one wall, and so on—while Peter felt nothing.
>
> So the way to state the rule is to say that *the man who has felt the accelerations*, who has seen things fall against the walls, and so on, is the one who would be the younger; that is the difference between them in an "absolute" sense, and it is certainly correct.[5]

Everything in this "explanation" is wrong.

Notice, first, that we were able to predict the effect without calculating the acceleration of anything: all we computed was the ratio of the lengths of the two trajectories. The accelerations play no role in explaining the end result. Indeed, it is a simple matter to alter the situation so that B is accelerated exactly as much, or even more than A, but still ends up older than A. In figure 11, we have added exactly matching accelerations to B's trajectory. Now B fires

[4] Rindler (1977), p. 46.
[5] Feynman, Leighton, and Sands (1975), vol. 1, p. 16-3.

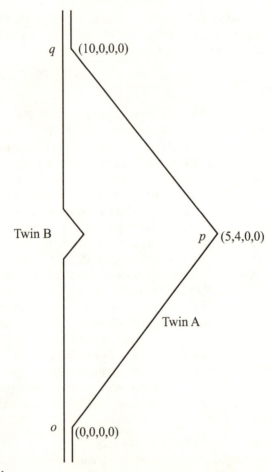

Fig. 11

his engines exactly to the same extent as A, but the little "bump" on his world-line will not significantly shorten it: he will still be the older of the two when they meet. By adding more such bumps, we can make B accelerate arbitrarily *more* than A and still be the older.

In Minkowski space-time, at least one of the twins must accelerate if they are to get back together: as mentioned above, a pair of straight lines in Minkowski space-time can meet at most once. This is incidental to the effect: in General Relativity, twins who are both on *inertial* trajectories at all times can meet more than once, and show differential aging when they meet. (We will see an example of this in chapter 6, p. 145.) Both Rindler and Feynman point out that acceleration is objective in Relativity, just as it is in Newtonian absolute space and time and in Galilean space-time. This is true but irrelevant: the issue is how *long* the world-lines are, not how *bent*.

Confusion 3 connects the Twins Paradox with "relativistic time dilation," which is portrayed as the thesis that "the faster a clock moves, the slower it ticks." Add this to the thesis that it is twin A who is "really moving" while twin B is "really at rest" (see Rindler and Feynman, above) and we have a perfect storm of misunderstanding. The Twins phenomenon is explained without having to attribute any "motion" or "speed" or "rest" to anyone: it is a simple matter of space-time geometry. Minkowski space-time, like Galilean space-time (and unlike Newtonian absolute space and time) supports no notion of "rest" as opposed to "motion." The trajectory of a body can be *straight* or *curved*, and hence *inertial* or *accelerated*, but there is no fact about "how fast a clock is moving." *A fortiori*, the effect cannot be explained by reference to how fast each twin moves.

The Law of Light, the Relativistic Law of Inertia, and the Clock Hypothesis all relate the behavior of perceptible items to the geometry of Minkowski space-time. To complete a relativistic dynamics, we would need a version of Newton's Second Law that relates the force put on an object to the curvature of its world-line. This requires more mathematics than we can afford to go into. But these three principles alone are sufficient to connect the geometry of Minkowski space-time to observable phenomena.

Minkowski Straightedge, Minkowski Compass

Special Relativity is, fundamentally, a postulate about the structure of space-time. But since the geometry of space-time is not immediately perceptible, the postulate becomes relevant to empirical predictions only via laws describing the behavior of perceptible things. Similarly, the supposed Euclidean structure of space can play a role in explaining the properties of visible diagrams only if the instruments responsible for the diagrams (straightedge and compass) reflect, and hence reveal, the underlying geometry of space itself. The nature of Minkowski space-time becomes more intuitive if we make this analogy explicit.

The role of a straightedge in Euclidean geometry is to indicate affine structure, that is, to pick out straight lines in the space. The Law of Light and the Relativistic Law of Inertia play exactly the same role in Special Relativity for a subclass of the straight lines in Minkowski space-time. In a vacuum, shielded from external influence, the trajectory of a light ray will be a straight, light-like line. Similarly, if shielded from outside influences, the trajectory of a massive body will be a straight time-like line. So our physical principles, framed in terms of the Minkowski geometry, entail that in certain circumstances, part of the affine structure can be made visible.

We have not offered any postulates that would help identify the straight space-like lines, lines that lie outside the light-cones of the events that compose them. These lines have a different physical character than the light-like and time-like lines, since (according to the Limiting Role of the Light-cone) they cannot serve as trajectories for physical entities. The basic difference between these sorts of lines should not come as a surprise: also in Galilean space-time, the straight lines that lie within a single simultaneity slice are fundamentally different sorts of things than the inertial trajectories through time. But as it turns out, we do not need to add to our physical postulates to identify the straight space-like lines: the behavior of light rays, inertially moving massive bodies, and clocks is sufficient to determine the whole geometry.

The Minkowski equivalent of a Euclidean compass would be a physical system that marks off events that all lie at the same Interval (along a straight line) from some central event. This requirement can be fulfilled if we accept the Clock Hypothesis. Suppose we have a large collection of identically constructed alarm clocks, all set to go off one minute after a button is pushed. At certain moment, the button is pushed and the clocks are sent off in all directions and

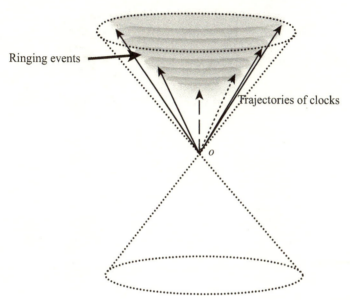

Ringing events

Trajectories of clocks

o

Fig. 12

with all different impulses from a central event (o in figure 12) and allowed to continue inertially. According to the Clock Hypothesis, the alarms will all go off when the length of their trajectories is 1 minute. What sort of shape will the ringing events make?

Let o be the origin of some Lorentz coordinate system, chosen so the Interval as defined in that system corresponds to minutes. Then the equation for identifying a ringing event p is

$$I(p,o) = \sqrt{T(p)^2 - X(p)^2 - Y(p)^2 - Z(p)^2} = 1.$$

That is, p will be a ringing event if it is in the future light-cone of o and if its coordinates (t, x, y, z) in this coordinate system satisfy $t_p^2 - x_p^2 - y_p^2 - z_p^2 = 1$. When we draw this locus of events on our Euclidean space-time diagram, the result is a *hyperboloid of revolution* (figure 12).

One stumbling block in interpreting figure 12 is that the points on the hyperboloid, which get unboundedly distant from o in the diagram, are all *exactly the same "distance" from o in the*

85

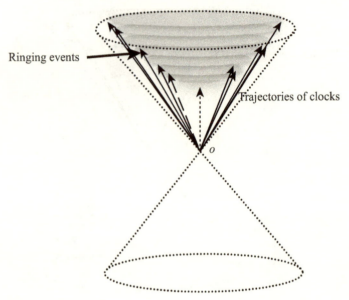

Ringing events

Trajectories of clocks

o

Fig. 13

intrinsic geometry of Minkowski space-time. Or, more accurately, the "lengths" of all the straight lines from *o* to the hyperboloid are all the same, even though they appear to be different lengths in the diagram. The fact that the "vertical" dashed trajectory seems "shortest" is purely a matter of convention: we are forced to arbitrarily depict one clock's trajectory as vertical, even though nothing differentiates it from any of the other clocks. Had we arbitrarily chosen the dotted trajectory on the right to be vertical in the diagram, the very same physical situation would be depicted as in figure 13.

It may help in interpreting figures 12 and 13 to compare them with figures 7 and 8. The "temporal length" of *any* trajectory that starts on one simultaneity slice and ends on the next is the same as any other, even though the lines on the diagram may appear to be of quite different lengths. Furthermore, the relationship between figures 7 and 8 is exactly analogous to the relationship between figures 12 and 13: in each case, both diagrams represent the same physical situation. So even though in figures 12 and 13 it appears that the "vertical" clock rings first and the others are delayed, there is nothing in the space-time geometry itself to support such a claim.

One essential difference between the temporal structure of Galilean space-time and Minkowski space-time is that in Galilean space-time, the elapsed time along a world-line is a function only of the simultaneity slices on which it begins and ends: further details of the trajectory are irrelevant. And without even mentioning simultaneity slices, we can say that in Galilean space-time, any pair of accurate clocks that start out together and end up together will show the same elapsed time between the two events. But as the Twins Paradox illustrates, this is not the case in Minkowski space-time.

Inertially moving bodies, light rays in a vacuum, and ideal clocks all can serve as instruments that make aspects of the geometry of Minkowski space-time visible. Using just these sorts of objects, we can pose and solve problems in Special Relativity. A few such problems have been provided in the appendix. These problems are designed to emphasize the fundamental geometry of Minkowski space-time and can be solved using nothing more than equation 2. (In particular, these problems are easily solved without use of the Lorentz transformations and so have a different character than most problems in physics texts.) There is no better way to come to appreciate the character of Minkowski space-time than solving problems like these, and these problems require only basic algebra. The reader might therefore benefit from pausing to peruse those problems now.

Using light rays in a vacuum and inertially moving clocks, we can also consider how to construct coordinate systems on our space-time. Since the clocks and the light are visible, these coordinate systems are not merely abstract: they have observational significance.

Chapter Four

Constructing Lorentz Coordinates

Our use of Lorentz coordinates has so far been entirely abstract: all we have asserted is that given the geometry of Minkowski space-time, there *exist* assignments of ordered quadruples of real numbers to events that have a specified relation to the intrinsic geometry of the space-time. We have not indicated how these coordinate systems might be practically realized. Indeed, without some principles connecting the geometry of space-time to the behavior of observable matter, no such practical realization could be described. Given the Law of Light, the Relativistic Law of Inertia, and the Clock Hypothesis, we are now in a position to indicate how to set up a visible coordinate system.

Our first task is to assign a *t*-coordinate to every event. We assume that we have available a limitless collection of ideal clocks and that we can tell that a clock is not subject to any external forces. All such clocks will, as per the Relativistic Law of Inertia, occupy straight time-like trajectories. *Arbitrarily* choose one such clock to be the "master clock" of the coordinate system. This arbitrary choice will determine many salient features of the resulting coordinates. The master clock will have some arbitrarily chosen time scale (seconds, let's say) and an arbitrarily chosen zero time. Given these, every event on the world-line of the master clock will be assigned a *t*-coordinate by the clock.

Extending the *t*-coordinate to events off the trajectory of the master clock requires making use of other instruments, including other clocks. But we must be very cautious about how these instruments are used. Given the sort of behavior displayed in the Twins Paradox, we cannot simply synchronize some auxiliary clocks at the master clock and then deploy them throughout space: the exact trajectories they take from the vicinity of the master clock to distant parts will influence their readings. So we assume rather that clocks are available on all the straight time-like trajectories in Minkowski space-time and set about identifying a collection of *co-moving* clocks.

Intuitively, two clocks are co-moving if they are both on inertial trajectories and are neither approaching each other nor receding from each other. It is important that the trajectories be inertial:

clocks rotating around a common center may maintain a constant separation, but they will not be considered co-moving. There is no analytical or conceptual guarantee that co-moving clocks are even possible: in some space-times, no pair of inertial trajectories maintains a constant separation. But the symmetries of Minkowski space-time allow for a rich structure of co-moving clocks.

An observer situated at the master clock can identify a co-moving inertial clock by radar ranging. That is, the observer sends out light rays from the master clock and then notes how long it takes (according to the master clock) for the light rays to be reflected off the target clock and return. It is especially simple to depict light rays on our diagram because the trajectory of a light ray in a vacuum is always represented by a straight line at a 45° angle in the diagram. If the target clock is co-moving, the round-trip time for the light will always be the same; if the target clock is in relative motion, then the round-trip times will change (figure 14). In order to make our space-time diagrams more legible, we will now restrict them to only two space-time dimensions, but the same technique would work in any direction.

By this means, the observer at the master clock can identify co-moving clocks. Target clock 1 is co-moving with the master clock, since every round-trip of light takes 2 minutes. Target clock 2, in contrast, is not co-moving: the first round-trip takes less than a minute and later ones more than a minute. In Minkowski space-time, a complete collection of co-moving clocks fills the entire space-time, partitioning it into a set of parallel straight time-like trajectories. Exactly one clock in the collection is present at each event, and our aim is to have the t-coordinate of that event be read off the appropriate clock. But before we can do this, we must calibrate and synchronize the co-moving clocks. Calibration, that is choosing a unit of measurement, is easy. Let the master clock emit a light pulse every minute, and then let the co-moving clocks adjust their calibrations so that one unit of time elapses between the receipt of successive signals. Now all the clocks are ticking "at the same rate."

The last step, synchronization, has received the most attention. In a classical space-time, such as Galilean space-time, it is clear what it means for clocks to be synchronized: they all assign

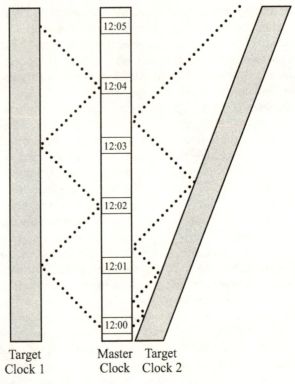

Target   Master Target
Clock 1   Clock   Clock 2

Fig. 14

the same time to events on the same simultaneity slice. Since in a classical space-time, there is an objective physical fact about which events happen together (i.e., at the same moment of absolute time), it is meaningful to demand that clocks be adjusted so they assign the same time coordinate to simultaneous events. But in Minkowski space-time, there is no such objective simultaneity structure for the clocks to reflect. There is no physical fact at all about whether two space-like separated events (i.e., events outside one another's light cones) "happen at the same instant." In this sense, the synchronization of clocks in Relativity requires a convention.

Given the geometry of Minkowski space-time, some conventions are simpler and more natural than others. In figure 14, observers can determine by radar ranging that target clock 1 is co-moving with the master clock while target clock 2 is not. They can further determine that the round-trip time for light to travel between the two clocks is always 2 minutes. It is extremely natural for target clock 1 to adjust its setting so that when a light ray arrives from the master clock, the target clock reads 1 minute later than the master clock read when the light ray was sent. Adopting this convention, target clock 1 will read 12:01 at the receipt of the first light ray, 12:03 at the receipt of the second, and so on. Adjusting every clock in the collection by this convention finishes our job: now every event in the history of the universe is assigned a t-coordinate. In figure 14, the set of events assigned the same t-coordinate in this coordinate system will form a horizontal line or plane, looking just like the simultaneity slices in figure 7. But since there is no physical relation of simultaneity among events, and since this particular way of assigning t-coordinates is arbitrary in many ways (including the choice of a master clock and the choice of a convention for adjusting the clocks), we will call these *equal-t slices* rather than simultaneity slices.

What if we had chosen a different master clock? In particular, what if we had chosen target clock 2 in figure 14 to serve as a master clock, identified co-moving clocks with respect to it, and then adjusted the calibration and setting of those clocks by the same recipe? The result is shown in figure 15, where target clock 1 has been synchronized to the master clock when they are adjacent at 12:00. What is of particular note is that the equal-t' slices in this new set of coordinates, which we will indicate by primed coordinates, are tilted with respect to the equal-t slices in the original system. In figure 15, the horizontal line indicates all events assigned the time 12:02 in the master clock system, and the tilted line all the events assigned the time 12:02 in target clock 2's system. If "coordinate simultaneity"—that is, being assigned the same t-coordinate in some coordinate system—were supposed to correspond to some real physical relation among events, at most one of these coordinate systems could be correct. But the non-existence of any objective simultaneity relation means that the

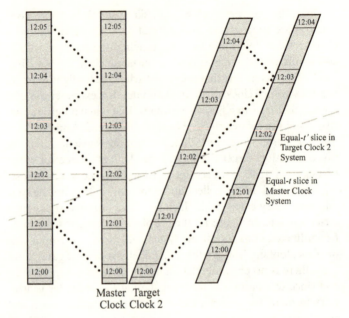

Fig. 15

t'-coordinates are no better or worse adapted to the objective geometry than the t-coordinates.

In our Euclidean space-time diagrams, the trajectories of co-moving clocks are related to their equal-t slices in the following way: as the trajectories of the clocks tip over in the diagram, the associated equal-t slices tip up so that the light rays always split the difference between the clock trajectories and the equal-t slices. In other words, if the slope of the master clock in the diagram is s, the slope of the equal-t surface for that clock's time coordinate is $1/s$. As a consequence, the angle in the diagram between a light ray and a clock trajectory always equals the angle between the light ray and the clock's equal-t slice. This is merely a consequence of the conventions used in drawing the diagrams.

The fact that the equal-t slices are different from the equal-t' slices goes by the name the *relativity of simultaneity*. It is commonly said that in Relativity, the notion of simultaneity is relative to an observer or to a state of motion. We can see the grain of truth in this characterization, but perhaps it also does as much harm as good. The key claim of Relativity is the *nonexistence* of simultaneity as a real physical relation among events. Furthermore, the "relativity of simultaneity" is a very derivative, coordinate-based circumstance. Different coordinate systems employ different *t*-coordinates, and their surfaces of constant *t*-value differ. The Lorentz coordinates are only a particular subset of all possible coordinate systems, distinguished by a certain simplicity of definition in Minkowski space-time. But in General Relativity, as we will see, Lorentz coordinates generally do not exist, and the practical procedures described above fail to define any coordinate system. (As mentioned, in General Relativity, co-moving inertial clocks generally don't exist, so our method fails at the first step.) The Twins phenomenon, in contrast, is a straightforward physical effect that can described without the use of any coordinate system and continues to exist in General Relativity.

Our two *t*-coordinates also illustrate *coordinate-based time dilation*. Relativistic time dilation is sometimes described by saying that "moving clocks tick slowly," but we already know that this locution makes no sense: in Relativity, there is no physical distinction between moving and nonmoving clocks, and all identically constructed ideal clocks, by definition, tick at exactly the same rate in proportion to the Interval along their world-lines. But if we examine the *t*-coordinate that results from using the master clock in our procedure compared with the *t'*-coordinate that results from using target clock 2 in its stead, we can see an interesting phenomenon. The master clock and target clock 2 are synchronized when they are adjacent at 12:00. But target clock 2 indicates 12:02 at a point *above* the 12:02 surface in the master clock system. In this sense, *relative to the master clock t-coordinates*, target clock 2 "runs slow." But the situation is perfectly symmetrical: the master clock also indicates 12:02 above the 12:02 surface in target clock 2's system of coordinates, so *relative to target clock 2's coordinates*, the master clock "runs slow." There is, of course, no fact about which clock is "really" running slow: this is a completely coordinate-dependent situation. In contrast, there is a fact about which twin in the Twins Paradox ends up younger, and that has nothing to do with coordinates at all. Mixing together the perfectly objective Twins phenomenon with the coordinate-based "time dilation" is a recipe for confusion.

So far, we have described a physical technique for assigning t-coordinates to all events in Minkowski space-time relative to an arbitrarily chosen inertial clock. To complete our coordinate system, we also need to assign x-, y-, and z-coordinates. This obviously demands an arbitrary choice of orthogonal x-, y-, and z-directions: we can imagine that our master clock has three orthogonal arrows attached to it in some way. The master clock must not be spinning, but, as Newton already noted, that can be checked empirically. The master clock serves as the spatial origin of our coordinates, so its x-, y-, and z-coordinates will all be 0. Every other co-moving clock must also be assigned some "spatial" coordinates. Let's consider how to do this along the x-axis of our coordinate system: the other axes will be treated the same way.

Since the master clock has an attached arrow pointing in the x-direction, it can send out light signals in that direction, which will be received by only some of the co-moving clocks. Those clocks will occupy the positive x-axis of our coordinate system: their y-and z-coordinates will be 0. Sending light in the opposite direction identifies the negative x-axis. The only question left is what x-coordinate value each of these clocks should be assigned.

The one piece of data available to the master clock is the round-trip time for a light signal to reach a co-moving clock and return. In figure 15, the round-trip time to the indicated co-moving clock is a constant 2 minutes. So a natural thing to do is to make the x-coordinate of that clock, which corresponds to its "spatial distance" from the master clock, proportional to that round-trip time. Even more simply, we can say that since the round-trip takes 2 minutes, the co-moving clock is 1 light-minute away. So the co-moving clock whose round-trip takes 2 minutes and which sits in the positive x-direction from the Master Clock get assigned x-coordinate $+1$, and the co-moving clock whose round trip takes 2 minutes and that sits in the negative x-direction gets assigned x-coordinate -1. Following such a scheme, every co-moving clock on the x-axis gets assigned an x-coordinate, and the other axes are dealt with in the same way. Repeating the procedure from the clocks on the axes ultimately results in every co-moving clock being assigned x-, y-, and z-coordinates, completing our coordinate system.

Lo and behold, by following this method, we have constructed a Lorentz coordinate system on Minkowski space-time. Different choices of master clocks, units of time, origin of time, and orientation of the spatial axes yield different Lorentz coordinates, and each set of Lorentz coordinates results from some such choice.

We should make the logical situation here clear. When we first introduced the notion of a Lorentz coordinate system, it was completely unconnected with any physical procedures: the coordinates were used only as an abstract way to specify the intrinsic geometry of Minkowski space-time. Next, we connected that geometry to the behavior of matter by a set of physical principles: the Law of Light, the Relativistic Law of Inertia, and the Clock Hypothesis. Finally, we have shown that if these principles are accepted, then a certain physical procedure, employing inertially moving ideal clocks and light rays in a vacuum, will result in the assignment of Lorentz coordinates to Minkowski space-time. At no point in this procedure have we so much as mentioned the "speed of light," or postulated that the "speed of light is constant": Minkowski space-time does not support any objective measure of the speed of anything. Nor have we anywhere invoked the notion of an "inertial coordinate system" or postulated that "all inertial systems are equivalent" or that "the laws of physics take the same form in all inertial systems." Rather, we have postulated a certain geometrical structure to space-time, invested that structure with physical significance for the behavior of visible matter by means of some physical postulates, and then described how to use the matter to construct coordinate systems.

Having done all this, we can now understand what is meant by the "constancy of the speed of light" and by an "inertial coordinate system" and by the "equivalence of all inertial systems."

Light, in itself, has no speed, since there is no absolute time or absolute space in Relativity. But *relative to a coordinate system*, we can assign a light ray a *coordinate speed*. For example, how "fast" does the light ray going from the master clock to the co-moving clock in figure 15 go? Well, *in terms of the unprimed coordinate system*, the t-value when it is emitted from the master clock is 12:00 and the t-value when it arrives at the co-moving clock is 12:01, so the difference in t-values (the "time elapsed" in this coordinate system) is 1 minute. Notice that the "time elapsed" is not measured by any individual clock: it depends on our procedures for adjusting the co-moving clocks. So in this coordinate-dependent sense, the light ray took 1 minute to get from the master clock to the co-moving clock. How "far" did the light ray go in making that trip? Again, there is no *objective* answer to this question: we don't have anything like Newton's persisting absolute space. But the x-coordinate of the co-moving clock in the master clock's coordinate system is −1, and its y- and z- coordinates are 0. So (using the Pythagorean equation) we can say the co-moving clock is 1 light-minute away. The light ray travels 1 light-minute (in these coordinates) in 1 minute (in these coordinates), so its *coordinate speed* in this reference system is 1 light-minute per minute.

Now, this whole calculation should strike the reader as a cheat. Given how we set about assigning the coordinates, *of course* the "coordinate speed" of light is 1 light-minute per minute: if you measure time in minutes and assign distances in light-minutes, then *by definition* a light ray will travel 1 light-minute per minute. And in this sense, the calculation *is* a cheat: the result was already baked into the procedure for assigning coordinates. But this is not a bad result. Since we have been trying to get rid of Newtonian absolute space and time, Newtonian absolute velocities have to go as well. This holds for light as much as for anything else. In this sense, the "constancy of the speed of light" cannot be a fundamental physical principle.

But we should not go too far in the other direction. The coordinate speed of light is constant in all Lorentz coordinate systems, and the coordinate speed of light is *not* constant in other coordinate systems that could be defined on Minkowski spacetime. (For example, just combine the t-coordinate of one Lorentz system with the x-, y-, and z-coordinates of another.) But the fact that Lorentz coordinates, with their relations to the behavior of light and clocks, are possible *at all* is not a matter of convention. The fact that all light emitted from an event (in a vacuum) propagates along a light-cone is not a matter of convention. The existence of co-moving clocks, as we have defined them, is not a matter of convention. The postulation of Minkowski space-time is a physical thesis, not a convention.

As a "spatial" analog to coordinate-based time dilation, Lorentz coordinates also display a *coordinate-based Lorentz-FitzGerald contraction*. Referring again to figure 15, we know that in the master clock's coordinates, the left-hand co-moving clock has constant x-, y-, and z-coordinates of -1, 0, and 0, respectively. So in these coordinates, the left-hand clock maintains a constant "spatial separation" of 1 light-minute from the master clock. By parity of reasoning, the right-hand clock maintains a constant separation of 1 light-minute from target clock 2 in the Lorentz frame associated with those clocks. But that leaves entirely open what the "spatial" separation between target clock 2 and its co-moving clock is *in the master clock's coordinates*. Clearly, these two clocks keep a constant separation: their world-lines are parallel. But it is not obvious just what that separation is in the master clock's reference frame.

We have enough information to solve this problem once we attribute a definite trajectory to target clock 2. Let's suppose that just like twin A on the first part of his voyage, target clock 2 travels inertially from the origin through the event with coordinates (5,4,0,0) in the master clock's coordinates. (The clock will read 12:05 rather than 5 because we started it at 12:00, but we will use the simpler coordinates.) So in the master clock's coordinates, target clock 2 goes 4 light-minutes in 5 minutes: it is moving at 80 percent of the speed of light. Using these Lorentz coordinates as Cartesian coordinates on the space-time diagram, we can convert our problem into one of Euclidean geometry. Since the y- and z-coordinates of all the clocks are always 0, we will leave them out of account. The trajectory of target clock 2 corresponds to the equation $x = \frac{4}{5}t$. That is, the events on the trajectory of target clock 2 all have x and t coordinates that solve this equation. The trajectory of the right-hand clock is similarly described by the equation $x = \frac{4}{5}t + x_0$, where x_0 is the x-value of the point where its trajectory meets the $t = 0$ axis. Note that these two equations describe lines of the same slope and hence parallel lines. The light ray that originates at the origin and moves to the right is described by the equation $x = t$. Since light rays are always represented by 45°

lines, their slopes are always $+1$ or -1. Since we have an equation for the right-hand clock and an equation for the light ray emitted from the origin, we can solve for the coordinates of the event where this light ray meets the right-most clock. Using $x = t$, we eliminate x from the other equation and obtain:

$$t = \tfrac{4}{5}t + x_0$$

$$\tfrac{1}{5}t = x_0$$

$$t = 5x_0$$

Since $x = t$, the t value of this event is also its x value. So the coordinates of the event where the light ray hits the clock are $(5x_0, 5x_0)$.

Now that we have an expression for the point where the light ray gets reflected, we can find an equation for the reflected ray. The trajectory of the light ray reflected back from the right-most clock to target clock 2 has a slope of -1 and originates at $(5x_0, 5x_0)$, so its equation is $x = -t + 10x_0$. (This is the only line with slope -1 that passes through $(5x_0, 5x_0)$.) The event where this light ray meets target clock 2 must satisfy both this equation and the equation for target clock 2. We must simultaneously solve the equations $x = -t + 10x_0$ and $x = \tfrac{4}{5}t$. Using the second equation to eliminate the x from the first, we derive

$$\tfrac{4}{5}t = -t + 10x_0$$

$$\tfrac{9}{5}t = 10x_0$$

$$t = \tfrac{50}{9}x_0.$$

The t-coordinate of the event where the return light ray meets target clock 2 is $\tfrac{50}{9}x_0$. Since we also know that $x = \tfrac{4}{5}t$ at that event, the coordinates are $(\tfrac{50}{9}x_0, \tfrac{40}{9}x_0)$. But we have one more piece of information. At the moment the return light ray hits target clock 2, that clock shows 2 minutes elapsed time from the origin. Since, by the Clock Hypothesis, this is a measure of the Interval along that trajectory, we have $I = \sqrt{\left(\tfrac{50}{9}x_0\right)^2 - \left(\tfrac{40}{9}x_0\right)^2} = 2$. Squaring both sides and solving for x_0:

$$4 = \tfrac{2500}{81}x_0^2 - \tfrac{1600}{81}x_0^2 = \tfrac{900}{81}x_0^2$$

$$2 = \tfrac{30}{9}x_0$$

$$x_0 = {}^{18}\!/_{30} = {}^3\!/_5.$$

The point where the right-most clock meets the $t = 0$ axis has co-ordinates $(0,{}^3\!/_5,0,0)$ in the master clock frame.

What this means is that *with respect to the master clock coordinates*, the two co-moving clocks are a constant ${}^3\!/_5$ light-minute apart, whereas in the coordinates of target clock 2 they are always 1 light-minute apart. And this circumstance is symmetric: in the coordinates of target clock 2, the master clock and its co-moving companion are only ${}^3\!/_5$ light-minute apart. This relation between different Lorentz coordinate systems is one way of understanding the so-called Lorentz-FitzGerald contraction.

The coordinate-based Lorentz-FitzGerald contraction is not, in any straightforward sense, the physical contraction of anything. In our example, all of the clocks always move inertially: nothing is subjected to any forces and nothing "shrinks." All we have noted is a fact about the *coordinates* assigned to various events in coordinate systems constructed by certain rules. This way of understanding the "effect" is encouraged, for example, by Rindler:

> According to Lorentz, the mechanism responsible for the contraction was a certain increment in electrical cohesive forces which tightened the atomic structure. Relativity, on the other hand, bypasses all explanation in terms of forces or the like, yet it predicts the phenomenon as inevitable. . . . In relativity the effect is essentially a geometric "projection" effect, quite analogous to looking at a *stationary* rod which is not parallel to the plane of the retina. Imparting a uniform velocity in relativity corresponds to making a pseudo-rotation in "spacetime."[6]

But just as there is a coordinate-based time dilation, which is not a physical effect, and the Twins phenomenon, which has nothing to do with coordinates, so too there is a coordinate-based Lorentz-FitzGerald contraction described by Rindler and also a *physical* Lorentz-Fitzgerald contraction that does have its explanation "in terms of forces or the like." Confounding these two

[6] Rindler (1977), p. 41.

"effects" can only breed confusion about Relativity and about experimental determinations of the "speed of light." We will devote the next chapter to clearing up this matter.

Before turning to that difficult topic, let's do one more simple problem that illustrates the relations between coordinate systems, and the role of our physical principles. Suppose we set up a pair of inertially moving markers that are, in any Lorentz coordinate system in which they are at rest, 1 light-minute apart. A clock, moving inertially, passes first one marker and then the other. *According to the clock*, exactly one minute passes between the event when it encounters the first marker and the event when it encounters the second marker. Questions: *In the reference frame of the markers*, how much time passes between these events? In this reference frame, how fast is the clock moving? *In the reference frame of the clock*, how far apart are the markers?

The easiest way to solve these problems is to draw a convenient space-time diagram. When trying to determine how a situation is described in some Lorentz coordinate system, the most convenient space-time diagram is one in which the constant t slices are horizontal and the constant x slices are vertical: these Lorentz coordinates on the space-time correspond to Cartesian coordinates on the diagram. If we make the event where the clock meets the first marker the origin of the coordinates, we get the diagram for the reference frame of the markers shown in figure 16. We want to calculate t_m, the t-coordinate assigned by the markers' Lorentz coordinates to the event where the clock meets marker 2.

We cannot jump to the tempting conclusion that if the clock travels 1 light-minute in 1 minute (as the clock itself shows), then it must be going at the speed of light. For the clock's readings are not the same as the t-coordinate in the marker frame. But the problem becomes trivial once we recall that an ideal clock—any ideal clock on any trajectory—measures the Interval along its world-line, and we recall how the Interval is expressed in terms of Lorentz coordinates. The Interval from $(0,0)$ to $(t_m,1)$, along a straight trajectory, is

$$\sqrt{(t_m - 0)^2 - (1 - 0)^2} = \sqrt{t_m^2 - 1}.$$

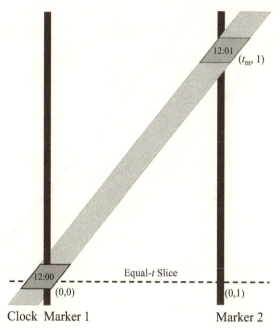

Fig. 16

As measured by the clock, we know that this same Interval is 1. So our complete calculation is

$$\sqrt{t_m^2 - 1} = 1$$

$$t_m^2 - 1 = 1$$

$$t_m^2 = 2$$

$$t_m = \sqrt{2}.$$

In the marker's reference frame, the clock takes about 1.414 minutes to travel between the markers, and so its coordinate speed in that frame is about 70.7 percent the speed of light. As judged from that frame, the "moving" clock "ticks slowly": it only registers

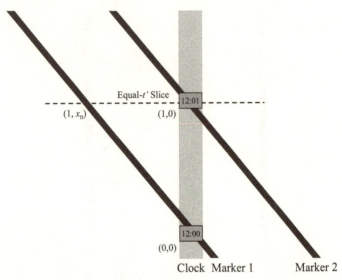

Fig. 17

that 1 minute has passed. This is the coordinate-dependent time dilation.

If we want to tell the same story from the point of view of the clock (i.e., in Lorentz coordinates in which the clock is at rest), then we should redraw the same space-time diagram making the clock's trajectory vertical (figure 17). The coordinates in figure 17 are now in the clock's Lorentz frame (i.e., the Lorentz frame in which the clock is at rest at the spatial origin) rather than the markers' frame. All of the events are exactly the same, but they receive different coordinate values. We would like to determine x_n, which is the constant coordinate distance between the markers in this frame.

The simplest derivation uses the results of our last calculation, together with the fact that if the velocity of A in the rest frame of B is $\mathbf{v}$, then the velocity of B in the rest frame of A is $-\mathbf{v}$. So if the clock is travelling at 70.7 percent of the speed of light in the rest frame of the markers, then the markers are travelling at 70.7 percent of the speed of light (in the opposite direction) in the rest frame of the clock. But then marker 1, 1 minute after it has passed the clock, will be at the event with coordinates $(1, -.707)$. That is, in the clock's coordinate system, the markers are always .707 light-minute apart. So we have derived the coordinate-based Lorentz-FitzGerald contraction.

Students of physics may note that we have derived all of these results without ever writing down the Lorentz transformation, that is, the general coordinate transformation between Lorentz coordinates that have the same origin and calibration. More examples of this sort of problem may be found in the appendix.

Before turning to the *physical* Lorentz-FitzGerald contraction, there is one final set of potential confusions to be addressed. We have been careful to pay attention to the use of coordinates in our descriptions: when we speak of the "frame of reference of the clock," we mean a set of Lorentz coordinates in which the clock's coordinate speed is zero. If we speak colloquially of "how things appear" in such a reference frame, we mean only how events are assigned coordinates in that system. This should not be confused with a very different sense of "how things appear to an observer," namely literally what the observer would *see* if she opens her eyes. That is determined by the light that comes through her pupils and has nothing at all to do with coordinates or reference systems.

For example, the coordinate-based time dilation describes how Lorentz coordinates associated with an observer in inertial motion will be attached to a clock that is moving with respect to that observer. In this sense, a "moving" clock will always "appear" to the observer to "tick slowly." But this does not at all describe what the observer will literally *see* if she uses a telescope to observe the clock. Literal observation in this sense (rather than ascription of coordinates) sometimes presents the clock as running slow and sometimes as running fast. If we return to the Twins scenario, it is easy to add light rays to the diagram and determine, without use of *any* coordinates, what each twin will literally see. Suppose, to make things simple, that each twin sends out a light pulse when he judges, by his clock, that 10 days have elapsed. These light pulses are shown in figure 18.

As the diagram indicates, the first flash sent by Twin B arrives at Twin A exactly at the turn-around event. (The flash is sent from

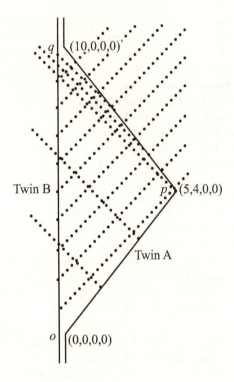

Fig. 18

the event with coordinates (1,0,0,0) and the slope of the trajectory is +1, so the flash reaches the point (5,4,0,0).) Since twin A has aged 30 days in the interim, if he has been keeping close watch on his brother throughout the trip, he has seen his brother and his brother's clocks moving very slowly, at ⅓ their normal rate. That is, in the course of 30 days (by his own clocks), twin A sees his brother go through 10 days of activity (by the brother's clocks). On the return trip home, the situation is reversed. Twin A receives 9 flashes while he experiences only 30 days of time: his brother, seen through the telescope, appears speeded up like the Keystone Kops. So in the 30-day return trip, twin A sees twin B age 90 days. At the end, his brother will be seen to have aged 100 days throughout the trip, as he must.

Twin B's experience is both similar and different. When he sees his brother receding from the earth, the brother also appears, to visual inspection, to be slowed down by a factor of three. Twin A's 30-day flash, sent out at the turnaround point, arrives at twin B after 90 days, by his reckoning. So for a full 90 days, twin B sees twin A aging slowly, appearing only 30 days older when he reverses his engines. On the return trip, twin A appears to twin B to age at triple speed. But twin B only sees the return trip for 10 of his days. In that period, twin A seems to age 30 days, and so he will be 60 days older when they reunite, as he must be.

There are no general rules about how fast a moving clock *appears* to tick if by "appears" we mean its literal visual appearance to an observer. The appearances can be determined by examining a space-time diagram, and there is no need to use any coordinate system at all. This is a physical question, and the introduction of coordinates is only for convenience of calculation.

The Physics of Measurement

THE CLOCK HYPOTHESIS

In order to connect the geometry of space-time to the observable behavior of perceptible items, we require some physical principles. In the last chapter, we investigated some of the observable consequences of the postulate that space-time has a Minkowski geometry, together with the Law of Light, the Relativistic Law of Inertia, and the Clock Hypothesis. We also assumed that we could determine that a body is subject to no external forces and that a particular region of space-time is a vacuum. Granting all of this, we derived the Twins phenomenon and showed how Lorentz coordinates could physically and visibly be constructed in space-time.

Of the three named physical principles, the Clock Hypothesis stands out as peculiar. The Law of Light and the Relativistic Law of Inertia describe how light rays and force-free massive bodies behave in Minkowski space-time, and they are couched in the sort of vocabulary we might expect in a physical law. But the Clock Hypothesis is a different matter altogether: it specifies how clocks behave (with respect to the Minkowski geometry), but "clock" is evidently not the sort of term that should appear in a fundamental physical law. Nature may recognize a distinction between light and massive particles, but Nature does not have to settle whether a given mechanism counts as a "clock" in order to determine how it should behave. A term like "clock," unlike "light ray" or "massive particle," cannot appear in the statement of any fundamental physical law.

Einstein was aware of this, and acknowledged that any discussion of physics couched in terms of "clocks" or "measuring rods" could be only a temporary expedient:

First, a remark concerning the theory [i.e., Special Relativity] as it is characterized above. One is struck [by the fact] that the theory (except for the four-dimensional space) introduces two kinds of physical things, i.e., (1) measuring rods and clocks, (2) all other things, e.g., the electromagnetic field, the material point, etc. This, in a certain sense, is inconsistent; strictly speaking, measuring rods and clocks would have to be represented as solutions of the basic equations (objects consisting of moving atomic configurations), not, as it were, as theoretically self-sufficient entities. However, the procedure justifies itself because it was clear from the very beginning that the postulates of the theory are not strong enough to deduce from them sufficiently complete equations for physical events sufficiently free from arbitrariness, in order to base upon such a foundation a theory of measuring rods and clocks. If one did not wish to forego a physical interpretation of the coordinates in general (something which, in itself, would be possible), it was better to permit such inconsistency—with the obligation, however, of eliminating it at a later stage of the theory. But one must not legalize the mentioned sin so far as to imagine that intervals are physical entities of a special type, intrinsically different from other physical variables. . . .[1]

We have avoided mention of "measuring rods" in our exposition in favor of just clocks and light rays, but the problem remains for clocks, and, as we will see, the issue of "rigid rods" will soon appear. What, then, is the exact status of the Clock Hypothesis, and how can we go about discharging Einstein's demand that clocks be nothing but moving atomic configurations, obeying the laws of electromagnetic fields, material particles, and so on?

The first thing to note is that the very term "Clock Hypothesis" is poorly chosen. It is a phrase used in the physics literature, but there is no way to state the content of such a hypothesis in a general way. Rather, what we should start with is not a hypothesis but a *definition*:

---

[1] Einstein (1982), pp. 59–61.

Clock Definition: An ideal clock is some observable physical device by means of which numbers can be assigned to
events on the device's world-line, such that the ratios of differences in the numbers are proportional to the ratios of Interval lengths of segments of the world-line that have those
events as endpoints.

So, for example, if an ideal clock somehow assigns the numbers
4, 6, and 10 to events p, q, and r on its world-line, then the ratio
of the length of the segment $\overline{pq}$ to the segment $\overline{qr}$ is 1:2, and so
on. This is just what we mean by "an ideal clock" in the context
of Relativity.[2]

Given the definition of an ideal clock, we may then propose
the *hypothesis* that some particular, specified physical system is an
ideal clock, or that it approximates an ideal clock to some degree
of accuracy in certain specified circumstances. Such a hypothesis
could be justified in different ways. For example, if we propose
that two different systems are both ideal clocks, we could keep
them side by side and see if the "time differences" they report
are proportional to one another. If not, they cannot both be ideal
clocks. We could argue that the best explanation for the existence
of different sorts of devices that are accepted as timepieces, and
agree in this way with one another, is that they are all (nearly)
ideal clocks: their agreement would be explained because they
are all accurately measuring the same objective physical quantity.
This sort of reasoning would justify treating fine Swiss timepieces
as (approximately) ideal clocks even if we could not give a clear
account, in fundamental terms, of how they work. But the gold
standard for justifying the belief that a particular device approximates an ideal clock, as defined, is to give a complete physical
analysis of its operation and to demonstrate that it satisfies the
definition. This, Einstein insists, is what any complete and consistent physical theory must ultimately be capable of providing.

Discharging this theoretical obligation for a mechanical
Swiss watch, or a quartz watch, or an atomic clock, requires a

---

[2] Charles Misner, Kip Thorne, and John Wheeler insist upon this point in their
classic *Gravitation* (1973), p. 393.

tremendous amount of detailed analysis: we have to be able to completely describe the device in the vocabulary of fundamental physics and specify the laws that govern all the parts. We are not in a position to carry out any such analysis. But we can carry out an analysis of a very simple sort of idealized clock in terms of the physical principles we have to hand, and we will learn much from the exercise. The clock I have in mind is a *light clock*.

A light clock consists of a ray of light that is reflected back and forth between two mirrors. We assume that there is a mechanism (and here again we shirk the duty to give a complete physical analysis) that can register when the light ray reaches one of the mirrors and "ticks" when this occurs. We also assume that some mechanism keeps count of the ticks. In this way, the device assigns integers to particular events along its world-line.

Einstein mentions this sort of clock in his discussion:

> The presupposition of the existence (in principle) of (ideal, viz., perfect) measuring rods and clocks is not independent of each other; since a lightsignal, which is reflected back and forth between the ends of a rigid rod, constitutes an ideal clock, provided that the postulate of the constancy of the light-velocity in vacuum does not lead to contradictions.[3]

Einstein specifies that the mirrors should be attached to the ends of a "rigid rod," but since that is a concept we have not yet tried to introduce, we will see how far we can get without it.

Consider a system that comprises two inertially moving mirrors on parallel trajectories in Minkowski space-time, a light ray reflected successively between the mirrors, and a counter that keeps track of the number of times the light ray reaches one of the mirrors. *So long as the mirrors are free of forces*, this system behaves as an ideal clock: the successive "ticks" of the light ray meeting the mirror mark off congruent segments on the world-line of the mirror. In essence, we can replace the master clock and target clock 1 in figure 14 with mirrors, then use the arrival of the light ray at one of the mirrors as defining a clock. We are not here concerned to explain *how we could know* that the trajectories of

---

[3] Einstein (1982), p. 55.

the mirrors are parallel to one another: we simply take that as part
of the physical specification of the system in question.

This device, as defined, is subject to a crippling restriction:
we can only prove that it functions as an ideal clock so long as
both mirrors are in inertial motion. To be an ideal clock, a system
should continue to measure the Interval no matter what sort of
trajectory it follows. But let's start with a simpler question: how
would this system behave if we set about to change it from one
inertial motion to another by suddenly accelerating it in some direction? In short, how does our system behave if we *boost* it from
its present inertial motion into another?

Suppose that we set about to accomplish this task by giving
each mirror an identical push or impulse. As of yet, our description of this intervention is not sufficiently precise: we also have
to specify the exact events where these impulses are applied. In
figure 19 we depict a pair of impulses that are applied *at events
with the same t-value in a Lorentz frame in which the mirrors are
initially at rest*. The counter is attached to the right-hand mirror, and the numbers produced by the device are indicated on the
diagram.

Both before and after the impulse, the system functions as
an ideal clock should: the trajectory between events 1 and 2 is
congruent to the trajectory between 2 and 3; and the trajectory
between 3 and 4 is congruent to that between 4 and 5. But what
about the relation between segments before and after the impulse? Is the segment between 2 and 3 congruent to the segment
between 3 and 4, for example?

If we specify the impulse more exactly, we can calculate the
answer. Adopt a Lorentz coordinate system in which event 3 is the
origin and the coordinates of event 2 are $(-1,0)$. In these coordinates, the Intervals between 1 and 2 and between 2 and 3 both have
measure 1. Suppose that after the impulse, both mirrors are moving at $\frac{4}{5}$ the speed of light in these coordinates. Then the equation
for the right-hand mirror after the push in these coordinates is
$x = \frac{4}{5}t$, and the equation for the left-hand mirror is $x = \frac{4}{5}t - \frac{1}{2}$.
(The mirrors must be $\frac{1}{2}$ of a "unit" apart since light takes one
"unit" to make a round-trip between them). The equation for the
light ray emitted at event 3 is $x = -t$, so the coordinates of event

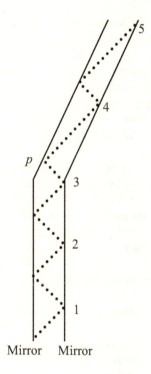

Fig. 19

p are $(\frac{5}{18}, \frac{-5}{18})$. (Solve $-t = \frac{4}{5}t - \frac{1}{2}$ to find this.) The equation for the light ray from p is $x = t - \frac{5}{9}$, so the coordinates of event 4 are easily found to be $(\frac{25}{9}, \frac{20}{9})$. The Interval between the event 3 and the event 4 is

$$\sqrt{\left(\frac{25}{9}\right)^2 - \left(\frac{20}{9}\right)^2} = \frac{5}{3}.$$

In sum, our light clock "ticks slower" after the impulse than before: successive later ticks correspond to a greater Interval than successive earlier ticks. Our system is not an ideal clock.

In order to continue to measure off the Interval at the same rate, the mirrors must somehow end up *closer together* after the impulse than they do in figure 19. To be precise, they must appear, in the original Lorentz frame, to be only $\frac{3}{5}$ as far apart as they were originally. So in addition to the force that sets the mirrors onto their new inertial trajectories, there must be some mechanism by which, as judged from this Lorentz frame, the mirrors are drawn closer together. Such a mechanism would produce a *physical* Lorentz-FitzGerald contraction. It is no coincidence that the degree of this physical contraction is exactly the degree of the coordinate-based Lorentz-FitzGerald contraction, but we still have no account of *the physical mechanism by which such a contraction is produced.*

Notice that the physical interactions we are about to consider are properly described as a *contraction* of the distance between the mirrors only from the point of view of some Lorentz frames, in particular from the point of view of the initial rest frame of the mirrors. In other Lorentz frames, the same change will rather be described as an *expansion* of the distance between the mirrors. But still, there must be some concrete physical account of the how the mirrors behave: we are doing physics here, not just analyzing coordinate systems.

We saw above that Rindler characterizes the Lorentz-FitzGerald contraction as merely a matter of geometrical projection rather than as a consequence of physical forces. This view is widely espoused by physicists, as was demonstrated by John Stewart Bell. Bell, who worked at the particle accelerator at CERN, reports the outcome of an informal experiment he conducted. He described to his colleagues a situation much like that of our two mirrors: two identically constructed rockets begin in inertial motion and at mutual rest. A signal is sent out so the rocket engines are started "at the same moment" as judged in their Lorentz rest frame. Since they are identically constructed, the trajectories of the rockets will be represented by congruent, parallel figures in a space-time diagram, so *as judged from the initial Lorentz frame,* the rockets maintain a constant distance from one another, just as our mirrors do in figure 19. Bell then added a small detail: there is a thread stretched tautly between projections on the sides of the two rockets. And Bell asked his colleagues, among the most respected theoretical and experimental physicists in the world, a simple physical question: will the thread break?

Bell reports:

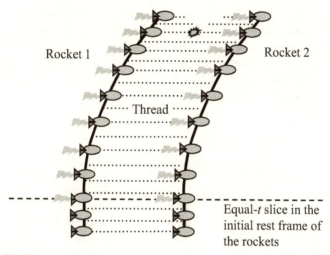

Rocket 1

Rocket 2

Thread

Equal-t slice in the
initial rest frame of
the rockets

Fig. 20

This old problem came up for discussion once in the CERN canteen. A distinguished experimental physicist refused to accept that the thread would break, and regarded my assertion, that indeed it would, as a personal misrepresentation of special relativity. We decided to appeal to the CERN Theory Division for arbitration, and made a (not very systematic) canvas of opinion in it. There emerged a clear consensus that the thread would **not** break![4]

But indeed the thread will break, as depicted in figure 20.

The physicists at CERN must have been thinking along the same lines as Rindler: if the Lorentz-FitzGerald contraction is merely a matter of *looking at the same events from a different angle*, or merely a matter of *describing the same events in a different coordinate system*, then of course it can't cause a thread to break! If it really has nothing to do with electrical or interatomic forces, then it can't have any observable physical effect. And indeed, what we have called the *coordinate-based Lorentz-FitzGerald contraction* is

[4] Bell (2008), p. 68.

nothing more than an observation about the relations between different Lorentz coordinate systems. But, as Bell rightly insists, there is also a *physical* Lorentz-FitzGerald contraction that *does* depend on interatomic forces and can have physical effects. We have to understand this physical effect in order to see how we could go about constructing an ideal clock.

Abstract Boosts and Physical Boosts

The coordinate-based Lorentz-FitzGerald contraction can be derived, as we have done, without anywhere discussing any physical item that is subjected to a force and hence accelerates. Even when we physically constructed Lorentz coordinates, we never had to deal with an accelerating body: all the clocks were always on inertial trajectories. So the coordinate-based contraction, by itself, can have no implications about the physical consequences of accelerating a system, such as the consequences of accelerating the mirrors in figure 19 or the rockets in figure 20.

In the physics literature, simply switching the coordinates from one Lorentz frame to a second in relative motion with respect to the first is sometimes called a *Lorentz boost*. We can ask how the same collection of events will be coordinatized under the two systems. Clearly, such a "boost" is an abstract descriptive change, not a physical change. It is often called a *passive boost*. On the other hand, we can keep to a single Lorentz coordinate system (or even use no coordinate system at all) and ask what would happen to a physical system if it were subject to a force and hence made to accelerate. The acceleration might be constant, as with Bell's rockets, or the result of a relatively sharp impulse, as with our mirrors. In the case of the mirrors, we take one and the same system from one inertial trajectory, before the impulse, to another inertial trajectory, after the impulse. This sort of *physical* change to a system can be called a *physical* or *active boost*.

Recall Galileo's ship from chapter 3. Galileo imagines doing a set of experiments in the ship while it is inertially "at rest" in the harbor. He then describes taking the *very same ship, containing the very same equipment* and letting it "proceed with any speed you like, so long as the motion is uniform and not fluctuating this way and that." That is, after it has come up to speed, the ship is to be again in inertial motion. So Galileo's ship corresponds to our mirrors in figure 19: it is on an inertial trajectory before and after a limited acceleration. And the physical claim that goes by the name "Galilean Relativity" is that *the observable outcomes of experiments will be the same before and after the acceleration.* Galileo makes no claims about what you would see *during* the acceleration: as the ship gets under sail the butterflies will indeed congregate at the stern of the ship, the water will not fall straight down, and so on. So Galileo's claim is that a system in inertial motion will behave the same way after a physical boost as before.

But our attempt at building a light-clock seems to refute this: the clock of figure 19 ticks at a different rate (in terms of measuring the interval) after the impulses are delivered than before. So does Special Relativity refute Galilean Relativity, or refute *the physical equivalence of all inertial frames*? Such outcomes would be exceedingly odd, since Galilean Relativity is often presented as a fundamental postulate of Special Relativity!

Looking back at figure 19, it should be obvious that the rate at which the clock ticks after the impulses are applied depends crucially on exactly *where and when* the impulses are delivered. If the impulse on the right had been delayed a bit, for example, then the mirrors would have ended up closer together, and the clock would have ticked faster (relative to the Interval). We chose to deliver the impulses at the same t-value in the initial rest frame of the mirrors, but that was a free choice. Perhaps we "ought" to have timed them differently.

In a real physical clock, this sort of question never arises, for a real physical clock will not contain two completely unconnected parts. In fact, the simple way to turn our pair of mirrors into a proper ideal clock is, as Einstein says, to *attach them both to a rigid rod*. Once we have done that, we don't have to worry about when we apply the pair of impulses: we can just push on *one* mirror and let the rod take care of bringing the other mirror along. During and shortly after the push, while the parts of the system are accelerating, the workings will go somewhat awry. But eventually, after we stop pushing, the system will come back to equilibrium in a new inertial trajectory. And if we do this with our mirrors of figure 19, we will find that, once everything settles down again, *the mirrors are closer together than is depicted in figure 19, and by exactly the right amount so the light clock ticks at the same rate after the boost as it did before.* Connecting the mirrors with a steel rod creates a much closer approximation to an ideal clock.

If we accept this result for the moment, we can see that the presence of the rod has a real, physical effect on the trajectories of the mirrors: the rod pulls the right-hand mirror a bit backward and the left-hand mirror a bit forward (as described in the initial rest frame). The rod produces these forces on the mirrors by a sort of real physical contraction, which is itself caused by the interatomic forces binding the rod together into a rigid body. In his article "How to Teach Special Relativity,"[5] which recounts the anecdote cited above, Bell produces a detailed physical analysis of this contraction for a system held together by electromagnetic forces and very gently accelerated. (The acceleration must be gentle or else the system will simply break apart rather than contract.) The thread in Bell's example breaks because these interatomic forces try to make it contract but the rockets prevent the contraction, building up tension until it exceeds the tensile strength of the thread. If we were to replace Bell's thread with a strong cable, then instead of the cable breaking, the tension in the cable will draw the rockets closer together, producing the physical contraction.

But if Bell has to do a detailed analysis of the electromagnetic forces in the atom to get his result, how can we be sure that other forces in other systems would produce the same seemingly fortuitous result, repositioning the mirrors in exactly the right way to maintain the proper functioning of the clock?

The key to a general analysis lies in the notion of a *rigid* body. Einstein mentions "measuring rods," and the essential characteristic of a measuring rod is that it is, in some sense, rigid. If we try to squash it, a rigid rod resists the compression, and if we try to stretch it, it resists as well. That is, a rigid body has an *equilibrium state* that it tends to maintain in the face of (sufficiently small) external forces, and it returns to that state when the external forces

[5] Bell (2008), chap. 9.

have been removed. Our simple two-mirror system, without the connecting rod, does not constitute a rigid system because the mirrors do not tend to resist external forces and maintain a fixed equilibrium state.

Complete physical understanding of an equilibrium state would require a complete account of the internal structure of the rigid system, both its composition and the forces among its parts. But even absent such a detailed account, we can make some general assertions about rigid bodies in any Special Relativistic theory. The fundamental requirement of a relativistic theory is that the physical laws should be specifiable using only the relativistic space-time geometry. For Special Relativity, this means in particular Minkowski space-time. It is the symmetry of Minkowski space-time that allows us to prove our general result.

Suppose a system has an equilibrium state that it tends to maintain when it is free of external forces (and hence in inertial motion). Specifying this state would typically require describing the precise internal structure and forces in the system. Let's call this equilibrium state S_{EQ}. In a given Lorentz coordinate system, such as the rest frame of the system, S_{EQ} will have a particular coordinate dependent description: for example, the relative positions of all the particles in the system can be given in terms of their coordinates in this Lorentz frame. Since S_{EQ} is an equilibrium state, the system will naturally tend to return to S_{EQ} if it happens to be in a state near to S_{EQ} and is released from external forces. If we deform the system a bit from S_{EQ} and then release it, it will spontaneously return to S_{EQ} as a consequence of its intrinsic structural forces.

Now consider a *different* physical state S' related to S_{EQ} as follows: S' has the same coordinate-based description relative to a *different* Lorentz coordinate system as S_{EQ} has relative to its rest frame. (Lorentz called such pairs *corresponding states*.) It follows, for *any* relativistic force laws, that S' will also be an equilibrium state, and that a system near the state S' and free from external forces will tend to go into the state S'. For the Minkowski geometry takes exactly the same form described in either Lorentz coordinate system (by the symmetry of Minkowski space-time), and the laws of physics take exactly the same coordinate-based form

117

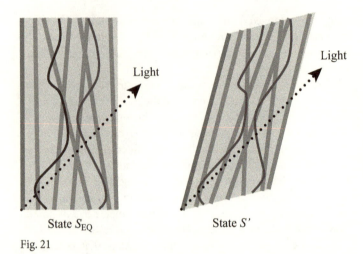

State S_{EQ} State S'

Fig. 21

when stated in a coordinate-based language in any Lorentz coordinate system (because the laws can only advert to the Minkowski geometry, and it has the same coordinate-based description). So the behavior of S' described in terms of the new Lorentz coordinates will be identical to the behavior of S_{EQ} described in terms of the old coordinates. The tendency of the original system to assume the state S_{EQ} if left on its own implies the tendency of the new system to assume state S'.

If S' is related to the new Lorentz coordinates in just the same way as S_{EQ} is related to the original Lorentz coordinates, it is easy to see how S' and S_{EQ} are related to each other, even if they happen to be quite complicated states. Figure 21 shows the situation schematically: in the diagram we just need to "skew" S_{EQ} in the right way to get S'.

Despite appearances, S_{EQ} and S' are geometrically congruent in Minkowski space-time. Generically, they are the same physical state. So if initially the system is disposed to return to S_{EQ}, after the appropriate physical boost it will be disposed to return to S'. Once again, the transition from S_{EQ} to S' must be sufficiently gentle: a pressed or stretched iron bar tends to return to its initial length unless it is pushed or pulled too hard, and deforms.

Bell's approach to the physical boost, which requires applying the appropriate dynamics to the accelerated system, is much more exacting than our generic observation. If Bell's program can be carried through, we can calculate exactly how the accelerated system behaves at all times and in all detail. For example, if we push on one side of our rigidified light clock, it will temporarily compress as the shock wave passes through it. During the acceleration, it will not behave exactly as an ideal clock should, and Bell's precise calculations would determine what these deviations from ideality would be. Our more generic approach has nothing to say about this: it only shows that once we stop accelerating the clock and allow it to seek out its new equilibrium state, it will again become an ideal clock ticking at its *original* rate with respect to the Interval. Furthermore, we do not even need to know the details of the forces that bind the clock together, only that the dynamical laws can be specified in terms of the Minkowski geometry. If these conditions hold, and we accelerate our rigidified light clock slowly enough, at the end of the experiment, it will have Lorentz-FitzGerald contracted itself to exactly the right degree to return to being ideal.[6]

It was mentioned above that the physical behavior described as the *contraction* of the thread in one frame will be described as the *expansion* of the thread in another. This follows from the symmetry of the Lorentz-FitzGerald contraction. Consider, for example, a Lorentz frame for figure 20 in which the firing of the

---

[6] The foregoing account of the physical Lorentz-FitzGerald contraction, which accounts for the breaking of the thread in Bell's example, traces the effect to three circumstances: the geometrical symmetries of Minkowski space-time, the necessity of specifying dynamical laws in terms of the Minkowski structure, and the physical constitution that makes the thread a rigid body (in the appropriate sense). Similarly for the account of why a light clock tied together by a rigid rod returns to the correct ticking rate after a physical boost. Harvey Brown (2006) has questioned whether facts about the geometry of space-time can play such an explanatory role, and he rejects the whole project of appealing to space-time geometry *per se* in physical explanations. But at the critical point in his book, where Brown seeks to cast doubt on the appropriateness of appealing to space-time geometry (and space-time itself) as a physical primitive in explanation, the objection is just this rhetorical flourish: "But is not this reasoning question-begging?" (p. 139). Brown does not specify exactly what question is being begged, so it is difficult to know how to respond. I take this account to demonstrate that no question is begged.

rockets causes the thread to *slow down* rather than to *speed up*. In such a frame, we would expect the interatomic structure of the thread to cause it to *expand* rather than to *contract*. But then why does the thread break?

In figure 20, the t-axis of this new reference frame will be parallel to the final trajectory of the rockets. That is, the new t-axis will be rotated clockwise relative to the old one. And as we have already seen, when the t-axis is rotated to clockwise, the corresponding x-axis rotates counterclockwise, so that the light rays continue to split the difference between the axes. The key now is to notice that in this new reference frame the rockets do not start to accelerate "at the same moment" (i.e., at the same t value). In this new frame, the rocket on the right begins accelerating to the right *before* the rocket on the left does. Small wonder that the thread breaks: as the rockets move apart (in this frame), the thread would have to stretch to accommodate the increasing distance between them. The thread is unable to stretch this much despite the fact that (in this frame) the interatomic forces are causing to thread to relax, rather than to tighten. In yet other frames, the rocket on the left starts accelerating earlier than the rocket on the right so the rockets are judged to get closer together, but the tightening of the interatomic bonds is so great that the thread cannot span even this reduced distance, and therefore breaks.

The surface contradiction between these three accounts of why the thread breaks illustrates that frame-dependent narrations of events in Relativity can be misleading. There is one set of events, governed by laws that are indifferent to which coordinate system might be used to describe a situation. In each frame-dependent account, the interatomic forces in the thread play a role in determining exactly when the thread breaks. But how that role is described in a particular reference frame depends critically on which frame is chosen.

THE "CONSTANCY OF THE SPEED OF LIGHT"

As we noted at the beginning of our discussion of Relativity, in order for any entity to have an objective, frame-independent speed, there must be an objective fact about how much time elapses between two events and how far apart in space the two events are. Once we abandon Newtonian absolute time and the persistence of points of Newtonian absolute space, there are no objective speeds, either of light or of anything else.

Given a coordinate system, we can define the coordinate speed of any object by treating one coordinate as playing the role of absolute time and the rest as playing the role of absolute space. For example, given Lorentz coordinates (t,x,y,z), we can define the coordinate time elapsed between events p and q as $t_p - t_q$ and the coordinate spatial distance between them as $\sqrt{(x_p - x_q)^2 + (y_p - y_q)^2 + (z_p - z_q)^2}$. Given these definitions, the coordinate speed of any ray of light in a vacuum is 1, since the Interval for such a ray is always 0. But this is a comment more about the definition of Lorentz coordinates than about the behavior of light.

Once we explain how to construct Lorentz frames by experimental procedures, the "constancy of the speed of light" takes on some empirical import. But since the constancy of the coordinate speed is built analytically into the definition of a Lorentz frame, the real empirical issue is just whether the constructive procedure will work at all. Furthermore, *no one ever actually assigns coordinates to events in a laboratory by the procedure discussed above.* The practical problem is obvious: the time required for a round-trip of light across a laboratory is too short to be reliably registered by a typical clock. So explicating the *empirical* import of the claim that the speed of light is constant by reference to the construction of Lorentz coordinates makes little contact with any actual laboratory operation ever carried out.

Nonetheless, there are results of actual laboratory operations that could naturally be described as indicating the constancy of the speed of light. These operations *do not employ clocks and measuring rods, or coordinates at all.* We are now in a position to understand why these experiments must come out a certain way if Special Relativity is correct. The physical Lorentz-FitzGerald contraction plays a critical role in these explanations.

Consider the following apparatus, which bears some similarity to an apparatus used by Hippolyte Fizeau to determine the speed

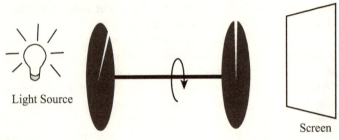

Light Source

Screen

Fig. 22

of light. Two disks are attached to a very rapidly spinning rod, each with a single radial slit cut in it. For light to pass through the whole apparatus, it must pass through both slits. Since the disks are spinning, the slits have to be offset by some angle for the light to get through, depending on the speed at which the disks spin, the distance between them, and the amount of time it takes for the light to get from one disk to the other (figure 22).

We could try to use the apparatus in figure 22 to determine the speed of light by measuring the distance between the disks, the speed of rotation, and the angle between the slits, but that would involve us in all the problems about measuring distance and time that we have discussed. Alternatively, we can simply *tune* the apparatus, by adjusting the speed of rotation or the angle between the slits or the distance between the disks, until light manages to get through both slits to the screen. Once the apparatus has been so tuned, there are various experiments that can be done.

We can, in the first place, vary the light source, including using sources in relative motion toward or away from the apparatus. If the Law of Light is correct, the light will still get through no matter the state of the source.

We can mount the apparatus on a turntable and rotate it so it points in different directions. This is similar to what Albert Michelson and Edward Morley did in their famous experiment, where the apparatus was mounted on a block of marble floating in a pool of mercury. Michelson and Morley used interferometry rather than our simple mechanical technique, but the idea was the same: the device could not determine the value of the speed of light (relative to the apparatus) but could be exquisitely sensitive to *changes* in the speed (relative to the apparatus). But no matter the orientation of the device or the time of year, the outcome never varied. The time of year would have been significant if the orbit of the earth changed its velocity relative to the supposed luminiferous ether.

Finally, we could put the apparatus on Galileo's ship, and test whether light produced on the ship and light produced on the shore manages to get through the device. These sorts of experiments would naturally be described (relying on Newtonian intuition) as "checking whether the speed of light is constant," even though they never involve actually determining what the speed of light is.

We now know that if (1) Special Relativity gives the correct account of space-time geometry, and (2) in all of the various configurations, the apparatus is in inertial motion,[7] and (3) the apparatus is a rigid body, then the outcome will never change: once properly tuned, light will get through the apparatus in all configurations. All three conditions listed above must hold for the result to follow. If we put the apparatus in Newton's bucket and spin it, for example, all bets are off. And when changing the setup from one where the apparatus points east-west to one where it points north-south, or from being at rest in the laboratory to being at rest in the (inertially moving) ship, the apparatus will have to be accelerated and allowed to settle back into its equilibrium state. What we have shown is that the symmetries of Minkowski space-time imply that these equilibrium states will be related to one another by Lorentz transformations. And therefore *any light ray passing through the device will bear the same geometrical relations to parts of the device in any of the experimental configurations.* Recalling again figure 21, we see that the light rays indicated on the diagrams bear the same geometrical relations to S_{EQ} as they do to S', because the light-cone itself is intrinsic to the space-time

[7] More exactly, if the apparatus is tuned when subject to some constant acceleration (including zero), it will give the same result when subject to the same acceleration.

geometry. So if a light ray manages to pass through a device in S_{EQ}, it will of necessity pass through a device in S'. This gives clear empirical import to the claim that "the speed of light is constant."

We must, however, always bear in mind the conditions under which these empirical predictions can be made. The apparatus must respond only to the *trajectory* of a light ray, not, say, its *color*. And the apparatus must be moving inertially when it is used. And it must be constructed of rigid materials and not be accelerated too strongly when changed from one inertial trajectory to another. Subject to these constraints, Special Relativity predicts that the outcomes of these experiments must always be the same, no matter how the apparatus is constructed. In this experimental sense, Special Relativity predicts, and explains, the constancy of the speed of light.

DEEPER ACCOUNTS OF PHYSICAL PRINCIPLES

Extracting experimental predictions from an account of space-time geometry requires some principles that connect the geometry to the observable behavior of physical entities. We have used three such principles in our discussion: the Law of Light, the Relativistic Law of Inertia, and the Clock Hypothesis. The first two of these, unlike the last, have the right form to be fundamental laws of nature. But they also might be consequences of deeper laws, couched in terms of different concepts.

For example, Maxwell's electrodynamics can be formulated in terms of Minkowski space-time: that is why considerations of electromagnetic phenomena led to Relativity in the first place. So we could postulate Maxwellian electrodynamics as a fundamentally relativistic physical law and then identify light rays as electromagnetic waves. We would then *derive* that the trajectory of a light ray in a vacuum will lie on the light-cone rather than postulating it.

Similar remarks apply to the Relativistic Law of Inertia. It is modeled on Newton's First Law of Motion and has the right form to be a fundamental natural law. But it might not be. We could imagine an account of the trajectories of massive bodies that has as a consequence that such bodies always, perhaps only approximately, follow straight lines through space-time when free of external forces. So long as these more fundamental laws are couched in terms of the Minkowski geometry, Special Relativity would be equally vindicated. The basic question is just whether the Minkowski space-time structure is all the space-time structure needed to do physics, and if the physical laws articulated in terms of that structure make accurate predictions. Maxwell's theory showed how this could be done for electromagnetism, and later work brought the Weak and Strong nuclear forces into the fold. But the attempt to produce Special Relativistic physics hit a snag with gravity. The upshot of this difficulty was the General Theory of Relativity.

General Relativity

Curved Space and Curved Space-Time

General Relativity was developed as a theory of gravity that incorporates the qualitative account of space-time structure found in Special Relativity. There are a few common myths about General Relativity that we need to dispel before presenting the theory. It is often said that General Relativity extends Special Relativity as follows: in Special Relativity all inertial reference frames (i.e., all Lorentz frames) are equivalent, and in General Relativity all frames of reference are equivalent; or in Special Relativity there is a physical distinction between accelerated and unaccelerated motion, but in General Relativity there is none; or in Special Relativity space-time has an inertial structure that is not a function of the distribution of masses, but in General Relativity it is a function of the distribution of masses (thus vindicating Mach). These claims are all false. General Relativity attributes an objective, intrinsic geometrical structure to space-time in exactly the same way that Special Relativity does, and that geometrical structure includes an affine structure that distinguishes accelerated from unaccelerated motion. The General Relativistic explanation of the phenomenon of Newton's rotating globes has exactly the same general form as the Special Relativistic explanation or, indeed, the explanation given in Galilean space-time: there is a tension in the cord connecting the globes exactly when the globes are accelerating (i.e., rotating). The acceleration is defined relative to the intrinsic structure of space-time: the globes might be the only material objects in existence. It is true that according to General Relativity, the geometry of space-time is *influenced* by the distribution of matter, but it is not *determined* by the distribution of matter. For example, in General Relativity, there are many distinct vacuum solutions, in each of which space-time is empty of all matter and energy. One of these vacuum solutions is Minkowski space-time.

It is also sometimes said that finding a relativistic theory of gravity was a particular challenge because the Newtonian gravitational force is instantaneous, while in Relativity there is no longer any notion of simultaneity by which an instantaneous action can be defined. This is wrong on several counts. First, it is very unlikely that Newton actually thought that the gravitational force is instantaneous: he thought that the force must be mediated by some sort of particle, which would have taken time to get from, for example, the sun to the earth. Of course, it was exactly here that Newton declared "*Hypotheses non fingo.*" But more critically, there is nothing in the general form of Newton's gravitational law that suggests difficulties for a relativistic version: Coulomb's law of electrostatics is an inverse-square force law, just like Newton's law of gravity, and Maxwell's electrodynamics returns Coulomb's law in the appropriate limit. But Maxwell's electrodynamics is fully relativistic. Indeed, not long after Einstein proposed Special Relativity in 1905, many different Special Relativistic theories of gravity were developed. Einstein himself worked on several but rejected them because they failed to satisfy either the Weak Equivalence Principle or the Strong Equivalence Principle, as we will soon see.[1] Only with the General Theory in 1915 were all of Einstein's criteria met, but the critical issue was not instantaneous action.

One more common claim about General Relativity is not quite accurate but does come close to the truth. This is the claim that in General Relativity, gravitational effects are explained by replacing the Euclidean space of Newtonian physics with a non-Euclidean "curved" space. The accurate claim is not that flat Euclidean space is replaced by a curved space but that flat Minkowski space-time is replaced by curved, non-Minkowski space-time. In fact, already in Special Relativity it is unclear what is meant by the *spatial* geometry of the world: just as there is no privileged or unique or objective notion of simultaneity of events in Special Relativity,

---

[1] See Norton (1992) for an extremely clear account of some of these attempts.

so too there is no privileged or unique or objective notion of "the geometry of space." A simple example can illustrate this.

Suppose that space-time has a Minkowski structure. What should we say about "the geometrical structure of space"? Since space is three-dimensional, the question can only be posed once we have selected some three-dimensional subspace in Minkowski space-time. Such a subspace should contain no time-like or light-like related events. But there are many such space-like subspaces that can be carved out of Minkowski space-time, and they display different geometrical structures.

The most obvious sort of "spatial" submanifolds of Minkowski space-time are associated with Lorentz coordinates. Given any Lorentz coordinate system (t,x,y,z), consider all of the events that have the same t value, that is, consider a "simultaneity slice" in that coordinate system. The Interval among events p and q is proportional to

$$\sqrt{\left(t_p - t_q\right)^2 - \left(x_p - x_q\right)^2 - \left(y_p - y_q\right)^2 - \left(z_p - z_q\right)^2}.$$

So if all the t-values of the events in the subspace are the same, the Interval among those events will be proportional to

$$\sqrt{-\left(x_p - x_q\right)^2 - \left(y_p - y_q\right)^2 - \left(z_p - z_q\right)^2} =$$
$$i\sqrt{\left(x_p - x_q\right)^2 + \left(y_p - y_q\right)^2 + \left(z_p - z_q\right)^2}.$$

Save for the factor of i, this has exactly the same form as the distance function in Euclidean space expressed in terms of Cartesian coordinates, and the factor of i washes out since we only care about the ratios among these numbers. So in a rigorous sense, the intrinsic geometry of an equal-t slice in a Lorentz coordinate system is three-dimensional and Euclidean.

But Minkowski space-time contains many other space-like surfaces. For example, the hyperboloid of figure 13 is another set of events in Minkowski space-time that one could call a "space," but its intrinsic geometry is not Euclidean. If one were to identify this set of events as "space," then one would say that in Special Relativity "space" is hyperbolic rather than Euclidean.

The study of non-Euclidean spatial geometry is a good preparation for understanding General Relativity, and non-Euclidean geometry is most easily grasped for two-dimensional spaces. We know that only Euclidean space admits Cartesian coordinates, so the investigation of non-Euclidean space requires that we forgo the luxury of simple coordinates. The best approach avoids coordinates altogether, and that is how we will proceed.

Euclidean space has many symmetries: it is homogeneous, isotropic, and scale-invariant. The simplest non-Euclidean spaces are homogeneous and isotropic and can described by simple axioms. The most familiar is the space of constant positive curvature: the surface of a sphere. On the surface of a sphere, the shortest line connecting two points is always an arc of a great circle, and these lines also satisfy an intuitive notion of straightness: given the command to walk straight ahead on the surface of the earth, one would naturally follow a great circle. If we identify the great circles as straight lines on the sphere and measure distances in the usual way, we will ascribe a non-Euclidean intrinsic geometry to that surface.

There are precise mathematical ways to characterize this geometry, but for our purposes it is sufficient to focus on some generic properties. Let's call a pair of straight lines in a two-dimensional space "locally parallel" if they both intersect some third straight line at right angles. In Euclidean space, locally parallel lines are really parallel: they never intersect. But on the surface of a sphere, locally parallel lines approach each other and eventually intersect, as illustrated by lines of longitude on a globe. All of these intersect the equator at right angles, and so are locally parallel, but they all meet at the north (and south) pole (figure 23).

It is obvious from figure 23 that the intrinsic geometry of the surface of a sphere is not Euclidean: the triangle formed by the two lines of longitude and the segment of the equator must have a sum of interior angles greater than two right angles since the angles at the equator alone are two right angles. The sum of interior angles of a triangle in this space is always greater than two right angles, the excess being greater in proportion to the area of the triangle.

In a hyperbolic space, or space of constant negative curvature, locally parallel lines behave oppositely: they diverge rather than converge. A good approximation to this space is given by a saddle-shaped two-dimensional surface in three-dimensional Euclidean

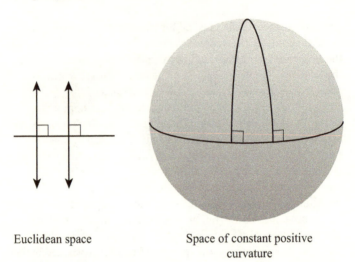

Euclidean space   Space of constant positive
                                              curvature

Fig. 23

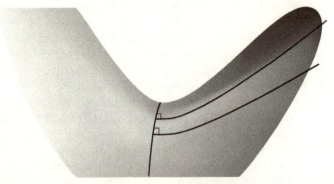

Fig. 24

space (figure 24). In this case, the sum of interior angles of any triangle is less than two right angles.

Beyond these especially simple spaces of constant curvature, there are spaces of variable curvature. These can be flat in some places, positively curved in others, and negatively curved in yet others. Or, the space may always have the same sort of curvature but different amounts in different places. Riemann developed methods for describing such spaces using general metric functions that can vary from place to place. The study of these various Riemannian spaces provides a helpful analog for understanding General Relativity. The basic analogy is this: as the Riemannian spaces of variable curvature are to flat Euclidean space, so are the space-times of General Relativity to flat Minkowski space-time. But to see how to employ this analogy, we first must consider some remarkable features of gravity.

GEOMETRIZING AWAY GRAVITY

If the most important experiment in the history of space-time physics was Newton's bucket experiment, the next most important was performed by Galileo. In *Two New Sciences*, Galileo sets out to refute Aristotle's claim that the speed at which bodies fall is in proportion to their weight. Since heavier bodies feel a greater force of gravity than lighter ones, it is natural to expect that the heavier should fall faster, but Galileo confirmed that the difference in speed of descent is negligible. If Aristotle had been correct, a body ten times heavier would fall ten times faster, but in the dialogue Sagredo reports:

> But I, Simplicio, who have made the test, assure you that a cannonball that weighs one hundred pounds (or two hundred, or even more) does not anticipate by even one span [i.e., about nine inches] the arrival on the ground of a musket ball of no more than half [an ounce], both coming from the height of two hundred brachia [about 300 feet].[2]

Galileo correctly assumed that the small difference in fall was due to air resistance, and that test bodies with different masses would fall at exactly the same rate if subject only to the force of gravity. This principle has come to be known as the *Weak Equivalence Principle*.

---

[2] Galileo (1974), p. 66. If Galileo's figures are accurate, he cannot be reporting experiments done at the Leaning Tower, since it is only half the height mentioned.

Newton's gravitational theory implies the Weak Equivalence Principle because it asserts that the force of gravity felt by a body is proportional to its mass. Mass, in Newton's account, plays three distinct physical roles: it is a measure of a body's natural resistance to being accelerated by a force (inertial mass), a measure of how much gravitational force a body produces on other bodies (active gravitational mass), and a measure of how much a body is affected by the gravity of other bodies (passive gravitational mass). The Weak Equivalence Principle, in this setting, asserts that the inertial mass of a body is proportional to its passive gravitational mass. If this holds, then the gravitational forces on two test bodies will be proportional to their inertial masses, so they will experience the same acceleration (if the only forces are gravitational). Hence the cannonball and the musket ball fall together.

Newton tested the Weak Equivalence Principle using pendulums made of different materials and found it to hold to one part in several thousand. Modern experiments have verified it to one part in 10^{12}. From a Newtonian perspective, the obvious explanation for the Weak Equivalence Principle is that inertial mass *just is* passive gravitational mass, so the proportionality must be exact.

Nothing vaguely like the Weak Equivalence Principle holds for forces other than gravity. For example, two test particles subject to only electromagnetic forces will accelerate in completely different ways if they are oppositely charged, or if one is charged and the other neutral. In this sense, the Weak Equivalence Principle says that every massive body couples to gravity in exactly the same way, unlike other forces whose effects on a body depend on how it is charged.

Figure 25 is a space-time diagram that illustrates Galileo's experiment as understood by Newton. Ignoring the motion of the earth, Galileo, at the top of the tower, is moving inertially. He feels a gravitational force downward, but it is exactly compensated by an equal and opposite force produced by the tower on the bottom of his feet. As soon as the balls are released, they are subject only to the gravitational force. The heavier ball is harder to accelerate, due to its larger inertial mass, but the gravitational force it feels is correspondingly larger, so it falls together with the lighter object. They fall side by side. Figure 25 also indicates the world-line of

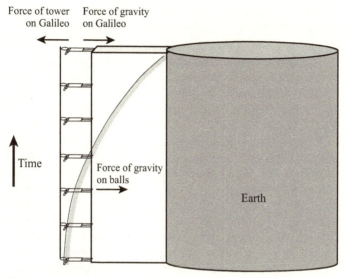

Fig. 25

the top of Galileo's head. Since the net force on his head is zero, this is a straight inertial trajectory, according to Newton.

There is no empirical problem with Newton's explanation of the Weak Equivalence Principle, but Einstein nonetheless detected a certain conceptual infelicity in Newton's explanation. In a way, we can trace Einstein's dissatisfaction back to Newton's own observations. The bucket experiment and the globes example relate the net force on an object to its absolute acceleration, that is, to the curvature of the object's world-line. And forces on an object can have observable sources, such as the tension in the globes' cord. Indeed, one way to make an "accelerometer" is to make the source of a force visible. For example, consider a simple weight free to move in a tube and attached by springs to the ends of the tube.[3] If we put such a device in a car with the tube oriented forward and step on the gas, the weight will "move backward" in the tube. The spring in front will stretch and the spring in back will compress

[3] This example is discussed by Nick Huggett in Huggett (1999), p. 136.

until they produce enough force on the weight to accelerate it together with the car. *Under the assumption that only the springs produce a forward-directed force on the weight*, the compression and tension in the springs measures the acceleration of the car. It is this sort phenomenon that Clarke had in mind when he wrote of the "shock" that a body experiences when brought from a state of absolute uniform motion to a state of absolute rest, that is, when it is accelerated.

Suppose Galileo stands at the top of the tower with such an accelerometer pointing downward. The weight will drop down the tube, stretching the top spring and compressing the bottom. According to Newton, this does not indicate any acceleration of the device: rather, the springs supply a force exactly sufficient to counteract gravity, and so keep the weight in inertial motion. Since the proper working of the accelerometer demands that only the springs exert force in the direction parallel to the tube, Galileo would be using the device incorrectly. And if Galileo releases the accelerometer and lets it fall, the springs will return the weight to the neutral position. Since the casing of the device falls at the same rate as the weight inside, no additional force is demanded of the springs to keep the two together. The falling accelerometer will not show any visible effect of acceleration: it will behave just as if it were moving inertially in empty space, with no gravity at all. Again, for Newton, this is a misuse of the device.

To arrive at the central concepts of General Relativity, all we need to do is to take this appearance—the failure of the falling accelerometer to register any acceleration—at face value. Returning to figure 25, suppose we simply delete the "force of gravity" from diagram. Then the "falling" balls have no forces on them at all, and Galileo has an *unbalanced* force on him in the direction away from the earth. Since the space-time version of the Law of Inertia says that objects with no forces on them follow straight trajectories through space-time, it would follow that the trajectories of the balls are straight, and the trajectory of Galileo's head is curved. The phenomena that we interpreted as showing the "equality of inertial and passive gravitational mass" are now understood in a completely different way. The two balls fall side by side because they are following essentially the same straight trajectory through space-time. *In General relativity, there is no "force of gravity"*: we mistakenly ascribe such a force to objects because we, like Galileo, are on *accelerated* trajectories when we are "at rest" on the earth. A reference frame attached to the surface of the earth is not an inertial frame, and the "force of gravity" used in such a frame is fictitious.

If we accept this picture, then we get the Weak Equivalence Principle for free: bodies that are subject "only" to gravity are really subject to no force at all, and so their trajectories will be the same. But in addition to the Weak Equivalence Principle, we also get a stronger result: the Strong Equivalence Principle. This states that the outcome of any experiment carried out "in free fall" in a uniform gravitational field will have the same result as if carried out in an inertial laboratory in empty space, and any experiment carried out "at rest" in a uniform gravitational field will have the same result as one carried out in a uniformly accelerating laboratory in empty space. For once we remove the force of gravity from figure 25, we see that Galileo, at the top of the tower, *is* in a constantly accelerated condition, so any experiments he does should reflect this fact.

The Strong Equivalence Principle is slightly fiddly. On the earth, the gravitational field (thought of in Newtonian terms) is not perfectly uniform: bodies do not tend to fall in parallel lines but fall rather slightly angled toward the center of the earth. They do not fall at a constant rate but fall faster the closer they are to the surface of the earth. So the Strong Equivalence Principle does not imply that one could not distinguish a laboratory "at rest" on the earth from a uniformly accelerated laboratory in empty space. In the latter case, bodies would "fall" (relative to a reference frame attached to the lab) at a constant rate in perfectly parallel lines. But so long as these slight inhomogeneities are disregarded, the outcomes of all experiments should be the same, because the physical situation in the two cases is the same: both labs are objectively accelerated. As another example, the trajectories of the two balls in figure 25 are not perfectly parallel: eventually the balls will collide as the both approach the center of the earth. These sorts of "tidal effects" can be important, but for the moment we disregard them.

One reason that the Strong Equivalence Principle is stronger than the Weak is that it has implications for *all* sorts of experiments. For example: one sends a light ray on an initially horizontal path from one side of a laboratory on the earth to the other. Will the light ray "bend," that is, end up closer to the floor when it hits the distant wall than it was when it began? Given only the Weak Equivalence Principle, there is no telling: the result would depend on *what light is*. If light is some sort of particle with inertial (and hence gravitational) mass, then it will be deflected downward. If it is some sort of wave phenomenon, or a massless particle, then the question is open.

But if the Strong Equivalence Principle holds, the light ray must "bend down." For in an accelerating lab in empty space, the floor of the laboratory will "accelerate upward" while the light is en route, so the light will end up nearer the floor. In short, the Strong Equivalence Principle predicts the "bending of light by a gravitational field" simply because the light is following the intrinsic geometry of space-time. The light is not "bent" at all: its trajectory is straight. Rather, the trajectories of the laboratory equipment, "at rest" on the earth, are bent. They are bent by the tangible forces put on them by the floor of the lab, the supporting tables, and so on.

One of the signal predictions of General Relativity—a prediction that distinguishes it from the various Special Relativistic theories of gravity extant in 1915—is the "bending" of light that passes close to the sun. Arthur Eddington's eclipse expedition of 1919 confirmed the effect; General Relativity has been the leading account of gravity ever since.

How does General Relativity explain gravitational phenomena in terms of the geometry of space-time? Once we accept that the trajectories of particles in free-fall are straight lines, it is clear that space-time cannot be Minkowskian. For example, consider the counterorbiting satellites of figure 26. According to General Relativity, the trajectories of the satellites are straight, but they also meet again and again.

Since no pair of straight lines in Minkowski space-time meets more than once, General Relativity must replace the flat Minkowski geometry with a curved alternative. And for exactly the same reason, our Euclidean space-time diagrams must be read with much

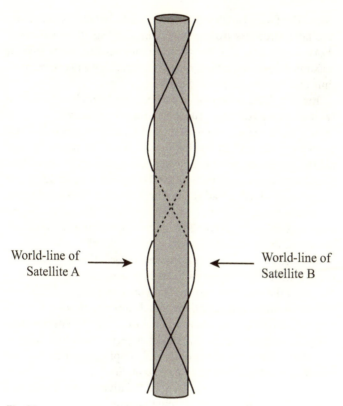

World-line of   →   ←   World-line of
Satellite A   Satellite B

Fig. 26

more care in General Relativity: we can no longer use only straight lines in the diagram to represent straight trajectories in the space-time. Neither the affine nor the metrical structure of the represented curved space-time can be easily read off the picture.

The straightness of the satellite world-lines follows from the decision to eliminate any force of gravity, but it also corresponds perfectly with the properties of straight time-like trajectories in Special Relativity. In Minkowski space-time, straight time-like curves have maximal length: a clock on such a trajectory will show more elapsed time between a pair of events than any other clock. The Twins Paradox gives an example. In General Relativity, free-fall trajectories are locally maximal: any small deviation will shorten their length. A straight trajectory need not be of globally maximal length, but a clock on such a trajectory will run off more time than nearby accelerated clocks.

Richard Feynman provided a wonderful example of the physical implications of these principles.[4] Suppose that there is a clock "at rest" at the base of Galileo's tower in figure 25. Since the clock is on a curved path in the space-time, it should not be recording the maximum possible time between events on its world-line. Take, for example, the event at which the clock reads 12:00 and the event at which it reads 12:01. There ought to be a way to have a clock begin at the first event also reading 12:00 and end at the second event having recorded a longer elapsed time. How can we maximize the time that runs off on such a clock?

If clocks on straight trajectories record (locally) maximal elapsed times, and if objects in free fall are on straight trajectories, then the answer is easy: at 12:00 the clock should be thrown straight upward with exactly as much force as needed for it to return to the base of the tower when the clock at the bottom reads 12:01. The second part of this clock's trajectory will look like the trajectory of the balls in figure 25, and, despite appearances in the diagram, the trajectory will be straight. So this is a straightforward empirical prediction of General Relativity—that the thrown clock will show more elapsed time than any other that starts and ends at these events—and it turns out to be correct.

The exact mathematical details of the space-time curvature go beyond our scope, but the form of Einstein's Field Equation can at least give the flavor of the theory.[5] The equation reads:

$$G_{ab} = R_{ab} - \frac{1}{2} R g_{ab} = 8\pi T_{ab}.$$

The quantity G_{ab} is called the *Einstein curvature tensor*. The first part of the equation gives the definition of this tensor in terms

---

[4] See, e.g., Feynman (1975) vol. 2, p. 42-12.

[5] See the suggested readings at the end of this chapter for resources that give proper, rigorous accounts of General Relativity. This account of the Einstein curvature tensor follows Baez and Bunn (2006).

of the space-time metric g_{ab}, the Ricci tensor R_{ab}, and the scalar curvature R. Suffice it to say that all of these are purely geometrical quantities. Physically, the Einstein curvature tensor at p determines how the volume of any small spherical swarm of particles in free fall at p, initially at rest with respect to one another, changes. If the Einstein curvature tensor is zero, then the volume of the swarm remains constant. (This is, of course, a comment about how the time-like geodesics near p behave: particles in free fall follow such geodesics.)

It is essential to distinguish the Einstein curvature tensor from the metric and from the full Riemann curvature tensor R_{abcd}. If the latter is everywhere zero, then the space-time is flat, and locally parallel geodesics neither converge nor diverge. But a zero Einstein tensor does not guarantee that the space-time is flat: the volume of a spherical swarm of particles can remain the same even though some particles converge, provided that other particles diverge. This behavior distorts the sphere into an ellipsoid, narrowing it in one direction and stretching it in an orthogonal direction. Because this sort of distortion of the oceans on the earth due to the gravitational influence of the moon causes the tides, it is called a *tidal effect*.

On the right-hand side of the equation, T_{ab} is the stress-energy tensor. It represents the distribution of matter and energy in the space. So the field equation essentially says that the way the volume of small balls of test particles in free fall behave in a region is determined by the amount of matter and energy in that region. The more matter and energy, the greater the Einstein curvature and the more the volume of the ball will shrink.

It is important to note that the Einstein curvature tensor does not describe the complete, detailed geometry of space-time. That job is performed by the metric g_{ab} (or R_{abcd}, which is interdefinable with g_{ab}) from which R_{ab} and R follow. So the complete distribution of matter and energy, T_{ab}, does not determine the space-time geometry but only constrains it. We have already seen an example of this: T_{ab} being uniformly zero is *consistent* with the space-time being Minkowski (i.e., flat) but also consistent with the existence of gravitational waves, which produce tidal effects. The distribution of matter and energy *constrains* the geometry of space-time but does not *determine* it.

At a large-scale, conceptual level, then, General Relativity overthrows some of the central aspects of Special Relativity. Special Relativity postulates a single, flat space-time structure, completely specifiable by the existence of global Lorentz coordinate systems. General Relativity implies that the spatio-temporal geometry of the universe depends on the distribution of matter and energy, and on further boundary conditions as well. In General Relativity, global Lorentz coordinates do not exist for any universe with matter, or even for most vacuum universes. One might suspect that all of Special Relativistic physics—electromagnetic theory, the theory of the nuclear forces, and so on—would have to be fundamentally modified to cohere with General Relativity.

This turns out not to be true. Just as the surface of the earth looks Euclidean in small patches—that is, the deviations from Euclidean geometry become small if the area considered is small—so the space-times of General Relativity look like Minkowski space-time in small enough regions. So if, like Tip O'Neill's politics, all physics is local (i.e., if the physics of a small region of space-time is determined completely by what happens only in that small region, by laws that make reference only to what is in that small region), then Special Relativistic theories can be used in a General Relativistic setting.[6] The deviations from Special Relativity only become manifest at a large scale, so appreciating the change from Special to General Relativity is easiest when we turn to astrophysics and cosmology. The next section considers two examples.

BLACK HOLES AND THE BIG BANG

Gravitational effects are intrinsically much weaker than the effects of the other forces of nature: the Strong and Weak nuclear forces and electromagnetism. We only notice the presence of gravity at all because it only works one way: all matter and energy

---

[6] Quantum mechanics seems to imply that not all physics *is* local, as we will see in volume 2. There are also much more detailed technical observations that can be made about how a Special Relativistic theory can be adapted to a General Relativistic setting. See Carroll (2004), sec. 4.7.

produces positive Einstein curvature, causing locally parallel free-fall trajectories to converge. Large conglomerations of matter, such as planets and stars, generate large Einstein curvatures. This is untrue of electromagnetism, for example, where the opposite influences of positive and negative charges can render large conglomerations of matter effectively neutral.

The natural dynamics of gravity therefore lead matter and energy to clump up. The focusing effect increases the density of matter, which in turn increases the Einstein curvature. Matter gets drawn together until some opposing force, such as pressure or the electromagnetic repulsion between atoms, produces equilibrium. This is what keeps all the matter in the earth from following inertial trajectories toward the center. If the accumulated matter becomes dense enough, electromagnetism can no longer keep the atoms accelerating off their inertial trajectories and the material will collapse further. More subtle quantum-mechanical effects can keep the increasing density at bay to a certain extent, so collapsing stars in a certain range of mass will stabilize again as neutron stars. But as we will see, beyond a certain level of density, the geometry of space-time itself dictates the outcome: nothing will stop the collapse. Such a situation produces a black hole.

The space-time geometry of a black hole provides insights into the milder situations of planets, stars, and neutron stars: far enough from the center, the black hole geometry resembles these. We do not have the mathematical resources to describe the black hole geometry in detail, so a qualitative space-time diagram will have to do. But all of our cautions about interpreting these diagrams must be redoubled. We can no longer draw the diagram so that straight trajectories correspond to straight lines, and Intervals bear no obvious relation to distances in the diagram. In fact, the only geometrical structure we will indicate in any detail is the light-cone structure. And even indicating this requires a choice of arbitrary pictorial conventions.

We will adopt the following convention:[7] the trajectory of light headed straight toward the center of the black hole will always be

---

[7] This convention is used, e.g., in Geroch (1978), chap. 8. This account is a truncated version of Geroch's.

represented by a line at a 45° angle to vertical, just as all light rays are depicted in our Minkowski diagrams. The lines representing light rays headed in other directions diverge from 45°. As a result, in the diagrams, the light cones appear to "tip" toward the center and also "narrow." This is purely an artifact of the representational conventions: "narrow" and "tipped" light cones are inherently just like any other light cones. The apparent tipping and narrowing is rather like the seemingly huge size of Greenland on a Mercator projection: the relative sizes on the map are not proportional to the actual areas represented.

Figure 27 represents the geometry. If there were a planet or star at the center, the light cones would gradually tip back up to vertical toward the center, but the black hole has a much more dramatic geometry.

The world-lines of four rockets are depicted: an exterior observer, who remains at a constant distance from the black hole, an explorer who departs from the exterior observer at *p* and goes into the black hole, and our two old friends Twin A and Twin B. The cylinder in the center is formed by the light rays that are vertical on the diagram, and it represents the *event horizon* of the black hole. The physical significance of the event horizon is particularly clear in this diagram: since the light cones become progressively more "tipped" as they get closer (in the diagram) to the singularity, no light that originates within the event horizon can ever escape it. Light can travel "vertically" along the event horizon, but any light ray that is properly inside the cylinder must eventually reach the singularity.

The behavior of light can be read directly off the diagram since the light cones are indicated. For example, the exterior observer emits a light ray at *q*, aiming it directly at the black hole. The explorer receives the light ray at *s*, meaning that she can still see the exterior observer when she turns around and looks in the opposite direction from the black hole. But when the explorer attempts to signal back, aiming a light ray backward at *s* toward the still-visible observer, we know she is bound to fail. Having passed the event horizon, her light cannot possibly make it back outside, and her light ray terminates at the singularity.

The explorer's attempts to communicate with the observer are still effective at *r*, as the diagram indicates. However, since

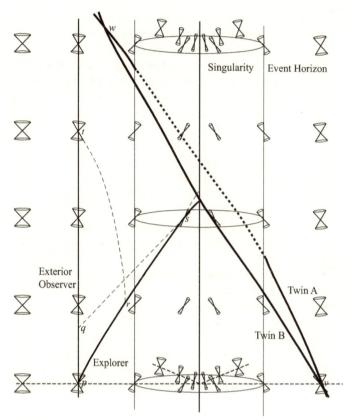

Singularity | Event Horizon

Exterior
Observer

Twin A

Twin B

Explorer

Fig. 27

the light cones have become "tipped," the trajectory of the light must start out nearly vertical on the diagram and can only slowly work its way back to the observer, who receives the message at *t*. Although one cannot directly read off the lengths of world-lines—and hence elapsed times—from the diagram, the general impression one gets is correct: as the explorer gets closer to the event horizon, the observer has to wait ever longer to receive messages from the explorer. Of course, a message sent at the horizon, aimed backward, will go "vertically" on the diagram and so never be received.

143

What happens to the explorer? The representation of her world-line terminates at the singularity in the diagram. The total elapsed time for her from p to the termination is finite: her clocks (if they continue to operate) never record more than a certain amount of time. It appears in the diagram that she "runs into" the singularity, but here the picture can be deceptive. In fact, the line marked "Singularity" in the diagram does not correspond to any part of the space-time represented at all. This should not be surprising: if there were actual physical events at the singularity, they should have future light-cones, but given how the narrowing light-cones are all converging from different directions, no smooth evolution to a single light-cone in the center would be possible.

In fact, the depiction of the singularity in the diagram can be highly misleading. To the incautious observer, the line in the center of the event horizon looks quite like the world-line of an object, as if the singularity were some sort of *thing* that abides at the very center of a black hole. One might even be tempted to regard it as some infinitely dense point, where all the mass that went in to form the black hole has accumulated. But notice that the singularity is not a *time-like* line at all: it cuts across the tops of the light-cones. Any massive object that approaches the singularity comes to have world-lines tilted at 45° in the diagram. Inside the event horizon, all "vertical lines" on the diagram are space-like rather than time-like. So "hitting the singularity" is not at all like colliding with some object, and one could not "avoid the singularity" by swerving around it, the way one can avoid hitting a tree. Rather, the singularity is an edge of space-time itself, where time-like curves simply cannot be continued. As Sean Carroll has remarked, one can no more avoid hitting the singularity than one can avoid hitting tomorrow: any maximally extended time-like curve inside the horizon must terminate there.

As the explorer approaches the singularity, gravitational tidal forces increase. Although the Einstein curvature is essentially zero along the world-line, locally parallel inertial trajectories are strongly focused in one direction and defocused in others. The explorer will become long and thin in the direction of the black hole, eventually being torn apart. This effect increases without limit, with the space-time curvature in particular directions growing without bound. And since Einstein's Field Equation cannot abide infinite curvature, there can be no solution at all at the points marked "Singularity" in the diagram. That is, in classical General Relativity, these points in the diagram represent no events at all. Space-time itself has come to an end.

In addition to the exterior observer and the explorer, we have depicted Twin A and Twin B in figure 27. We here redeem a claim made back in chapter 4: the Twins phenomenon can occur in General Relativity with no acceleration at all. Twin A and Twin B synchronize their clocks at event v. Despite appearances in the diagram, both of their trajectories are perfectly straight: the "bending" is due to the curvature of space-time itself. By means of this gravitational lensing, the twins meet again at event w, and one will typically be older than the other. The lengths of the two straight trajectories from v to w will generally not be the same: which of them is longer depends on the exact details. This example should put to rest the idea that the Twins phenomenon has anything to do with acceleration.

The idea that space-time ends with the singularity in a black hole naturally suggests the opposite possibility: space-time could also begin with a singularity. And indeed, if one retrodicts from the present expansion of the universe, one arrives at an early universe that grows in density as one goes back in time. Assuming General Relativity to hold at all energy scales, the result is the Big Bang: an initial singularity in the history of the universe.

Of course, the postulation of singularities associated with black holes and the Big Bang rely on unwavering commitment to the Einstein Field Equation. If General Relativity breaks down at high energy or high curvature, then all bets are off. Furthermore, these models postulate that the distribution of matter and energy in the universe can be exactly represented by a stress-energy tensor. It is unclear how this supposition coheres with the quantum-mechanical account of matter.

The problem of incorporating the quantum-mechanical account of matter into a theory like General Relativity is particularly acute because while the stress-energy tensor represents the local distribution of matter and energy in space-time, it is extremely unclear what sort of locally defined matter or energy distribution there is in a quantum-mechanical theory. This poses conceptual problems for understanding quantum theory that run much deeper than just reconciling it with Relativity. We will discuss these problems in volume 2.

The notion that space-time itself may have a beginning or an end is the most philosophically intriguing implication of General Relativity, but also the most tenuous. Many physicists are pursuing theories according to which the Big Bang is only the earliest phase in our local neighborhood of the universe, not the beginning of all existence. Similar speculations suggest that the explorer's trip need not end at a singularity, although it seems likely that the tidal forces would be fatal in any case. But any rigorous treatment of these topics requires a more advanced physics than we currently possess.

THE HOLE ARGUMENT

Back in chapters 2 and 3, we examined a pair of arguments offered by Leibniz against Newtonian absolute space and absolute time. Each of these arguments is founded on a symmetry in Newton's physics. The so-called *static shift* argument uses the translational and rotational symmetry of Euclidean space to argue that for every possible distribution of matter, the distribution generated by moving everything a fixed distance in some direction, or rotating everything a fixed angle around some axis, is also a possible distribution. If the original distribution fails itself to display the symmetry, then the new distribution is a distinct, physically possible state of affairs, consistent with all of Newton's laws. Similarly, the *kinematic shift* changes the absolute velocities of all matter by a fixed amount, again yielding a distinct physical possibility. Accepting these pairs as distinct possibilities was supposed to violate both the Principle of Identity of Indiscernibles and the Principle of Sufficient Reason, the latter with respect to God's choice to create one or the other of the possibilities. We noted in those chapters that the Principle of Identity of Indiscernibles has nothing to recommend it, and that the injection of theological considerations has no place in a physical argument.

The kinematic, but not the static, shift also illustrates that Newton was committed to particular questions of physical fact that could not be answered by observation. In particular, the exact magnitude and direction of the absolute velocity of any parcel of matter could not be determined by experiment, as is illustrated by Galileo's ship. But once we abandon Newton's absolute space and time for either Galilean or Special Relativistic space-time, that awkwardness vanishes: there are no longer any absolute velocities whose magnitudes and directions are inaccessible to us.

Since both the static and kinematic shifts appeal to symmetries of the space-time, one would expect these sorts of issues not to arise in General Relativity. A generic General Relativistic space-time simply has no such symmetries. Since the curvature of the space-time varies from place to place with the energy density, a generic solution to Einstein's Field Equation is neither geometrically homogeneous nor isotropic. The static and kinematic shift operations cannot be defined.

Nonetheless, John Earman and John Norton have put forward an argument (based on a somewhat different set of considerations articulated by Einstein) that a related and more radical problem infects General Relativity. They have called this the *hole argument*.[8] To appreciate the argument, recall the three sorts of transformation depicted in figure 1. The Leibniz-shift arguments employ isometries: the relative distances, and hence relative velocities, of objects are preserved under the transformation, while the metrical structure of the background space-time remains fixed. The hole argument, in contrast, employs a topological transformation. In particular, it employs a *hole diffeomorphism*: a smooth topological transformation that is the identity map outside some region (the "hole") but not the identity inside. Figure 28 illustrates such a hole diffeomorphism. One must understand the two diagrams as drawn on *the very same* background manifold, so that the locations marked *p* and *q* in the two diagrams are supposed to represent *the very same events*. All that has changed is how the matter and fields are distributed on those events.

---

[8] Earman and Norton (1987); Earman (1989), chap. 9.

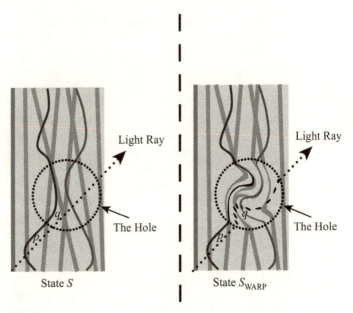

Fig. 28

At first glance, it may seem that if state S is a solution to Einstein's Field Equations, state S_{WARP} cannot be. For example, we know that a light ray in a vacuum ought to have a straight worldline, but the light ray in S_{WARP} seems highly curved. The trick to the hole diffeomorphism is that when we move things around according to the topological transformation, we move not just the matter and light *but the space-time metric as well.* The space-time metric is treated as the same sort of thing as an electromagnetic field, and so it gets transported by the diffeomorphism. If the dotted line in S is straight according to the original metric, then the dotted line in S_{WARP} is straight according to the transformed metric. Since the Field Equation relates the behavior of matter to the space-time metric, when we transform both the matter and the metric together, the new object is still a solution to the equation.

In order for S and S_{WARP} to represent different possible physical situations, there must be some physical difference in what they each represent. Such a difference can apparently be specified if we accept that p and q in each diagram represent the very same events. For in S, p and q are light-like separated, while in S_{WARP} they are space-like separated (note how the light ray divides them from one another in S_{WARP} but not in S). It is essential for the hole argument that the background topological space be considered "fixed" while both the matter and the metric are "moved around" on it between the two situations. The hole argument proceeds from the proposition that S and S_{WARP} each represent a physically possible situation, a solution to the fundamental physical laws, and further that these two situations are physically distinct.

If we grant this proposition, the argument then takes a different form than either of the Leibniz-shift arguments. We have already seen that the most damaging possible consequence of a Leibniz-shift argument is the existence of empirically inaccessible physical facts, such as the absolute velocity of an object. The hole argument does not seem to entail any such empirically inaccessible facts. For example, one can't say that it is empirically inaccessible whether the events labeled p and q are, *in fact*, light-like separated or space-like separated: just send light from p and see if it reaches q. For the singular terms "p" and "q" cannot just magically come to denote particular events: in order to attach such labels to particular space-time events, we would need some means to designate these events as opposed to others. And if we can pick out p and q at all, we can tell (in principle) whether they are time-like, space-like, or light-like separated.

The bite of the hole argument does not have to do with empirically inaccessible facts: rather, it is a claim about *determinism*. If we accept that S and S_{WARP} represent physically possible but ontologically different situations, then the problem arises because *outside the hole, the physical situations are identical*. It is critical to the argument that the hole diffeomorphism leaves the exterior of the hole unchanged. In this respect, the hole diffeomorphism is unlike the Leibniz-shifts: each of these alters the disposition of matter through all time. But if the exterior of the hole is exactly the same in each case, and if both S and S_{WARP} are physically possible solutions to the Field Equations, then the Field Equations themselves must be radically indeterministic: the entire physical situation outside the hole together with the laws of physics do not determine the physical situation inside the hole. Since the hole can be of any size and shape, and the warping can take infinitely many different forms, this spells indeterminism of the most radical stripe.

The heart of the hole argument is evidently a deeply metaphysical proposition, namely, that S and S_{WARP} can be interpreted as representing two metaphysically distinct and yet physically possible situations. Since the differences between S and S_{WARP} can only be stated in terms of individual, named events (rather than qualitative features, such as the existence of a pair of colliding particles, or a black hole), the metaphysical issue of the identity of individuals in different "possible worlds" must be addressed. One line of reply to the hole argument, presented by Jeremy Butterfield (1989), argues that S and S_{WARP} (which are merely mathematical representations) cannot represent *metaphysically distinct* situations, since an event in S can only be identified with an event in S_{WARP} by means of some counterpart relation. But any counterpart relation will match up the event p in S with the event that the diffeomorphism maps p to in S_{WARP}, and the same for q. So if p and q are light-like related in S, their counterparts are light-like related in S_{WARP}. According to David Lewis's semantics for modal locutions, this means that if p and q are actually light-like related, there is no possible world according to which they are not light-like related.[9]

If one adopts Saul Kripke's approach to modal semantics, a different solution presents itself. Kripke insists that cross-world identities need not be determined by counterpart relations: they can be directly stipulated when describing a situation.[10] Thus, a particular point in S_{WARP} represents the actual event p simply because I say that it does when constructing S_{WARP}. Suppose S correctly represents the actual world, and in the actual world I name two events p and q, respectively. Then S_{WARP} does indeed represent the events p and q as space-like related rather than as light-like related (as they actually are). Contrary to Butterfield's approach, S and S_{WARP} repre-

---

[9] See, e.g., Lewis (1986), chap. 4.
[10] Kripke (1980), pp. 17, 44.

sent *distinct* situations. But it is not at all clear that S_{WARP} represents a physically or metaphysically *possible* situation. I can say "If Nixon were a ham sandwich . . ." and thereby produce a representation according to which Nixon was a ham sandwich, but it does not follow that in any sense it is *possible* for Nixon to have been a ham sandwich. It is contrary to the essential nature of the actual Nixon that he could have been a ham sandwich, all of my linguistic stipulations notwithstanding. Similarly, we can argue that if the particular actual individual events p and q are light-like related, then it is not metaphysically or physically possible for *those very events* to have been space-like related: the spatio-temporal relations among space-time events are essential to their identity. This line of response can be found in Maudlin (1989).

Yet a third response, offered by Carl Hoefer and Nancy Cartwright (1994), grants the radical indeterminism that Earman and Norton propose but insists that such indeterminism is harmless and unworrisome. S and S_{WARP}, if they are different possible physical situations, will look and feel exactly the same to any creature that happens to exist in them. No describable experiment can have an observably different outcome in S than it has in S_{WARP}. Just as with the static shift, no intelligible question about the actual world will go unanswered due to empirical inaccessibility. Earman and Norton think that the indeterminism is still unacceptable, even if it has no empirical consequences, because it is not derived from any physically plausible source. Determinism, they say, might be false but it ought at least to have a fighting chance. And indeed, it should be clear that the basic form of the hole argument would apply in Galilean and Newtonian space-time as surely as in General Relativity: if space-time events only have their geometrical structure contingently, then the hole construction will work in any space-time.

There are, then, multiple lines of response to the hole argument.[11] Still, the argument might inspire some radically new metaphysical account of space and time, an account that is immune from this form of argument. To date, no such alternative has arisen. Only time will tell.

---

[11] A more complete discussion of the literature can be found in Norton (2008).

Chapter Six

The mathematics of General Relativity is complex, and this chapter gives only a vague and qualitative overview. There are texts at all different levels that present more details. Geroch (1978) is a good place to start for discussion at a very mild level of technical detail. Notable among physics texts that emphasize geometrical understanding are Carroll (2004) and the online notes of Baez and Bunn (2006), as well as the Track 1 sections of Misner, Thorne, and Wheeler (1983). Other nice texts are Wald (1984) and, at a heavily mathematical level, Hawking and Ellis (1973). Texts by philosophers that present and analyze the theory in detail include Friedman (1986) and Earman (1989, 1995).

The Direction and Topology of Time

THE GEOMETRY OF TIME

Time is one-dimensional. The exact import of this seeming triviality is easy to understand if one accepts Newton's account of absolute time. Absolute time is made up of instants or moments, each of which contains infinitely many events that happen simultaneously. In an essay called "On the Gravity and Equilibrium of Bodies," in the course of making an analogy, Newton employed a striking description of an instant of time and the relation between time and space:

> For we do not ascribe various durations to the different parts of space, but say that all endure together. The moment of duration is the same at Rome and at London, on the Earth and on the stars, and throughout all the heavens. And just as we understand any moment of duration to be diffused throughout all spaces, according to its kind, without any thought of its parts, so it is not more contradictory that Mind also, according to its kind, can be diffused throughout space without any thought of its parts.[1]

For Newton, a single moment of time embraces all of space, and absolute time is the totality of all such moments. To say that absolute time is one-dimensional for Newton, then, is just to say that these moments have a one-dimensional geometry: a "time line." What is characteristic about a one-dimensional geometry, as opposed to a geometry of more dimensions, is that at any point there are (at most) only two directions: left and right, say, or up

[1] The essay "De Gravitatione" can be found in Newton (1962) and is excerpted in Huggett (1999), p. 113.

and down, or forward and back. Similarly, given any moment of Newtonian absolute time, there are only two temporal directions from that moment: toward the future and toward the past.

To say that something is one-dimensional in this sense does not fix all of its geometry. The equator is one-dimensional, and at any point there are only two directions one can move while staying on the equator: to the east and to the west. But the equator has the further feature that proceeding far enough to the east, or to the west, brings one eventually back to the starting point. Given any point p on the equator, the rest of the points do not divide into those that are to the east of p and those that are to the west of p: every point (including p itself) can be reached by a trip that goes in either direction. This is a fact not about the dimensionality of the equator but about its topology.

Newton assumed without comment that the topology of time is different: if one goes on indefinitely into the future, one will never come again to the present time. The past, we say, is over and done with: it will never come again. The topology of time, according to both Newton and to common sense, is not like that of the equator: any event that is not happening now lies either to the future or to the past, but not both. The universal time line, for Newton, is not a circle.

Newton's account of absolute time relies on the notion of absolute simultaneity. Instants of time are "diffused throughout all spaces" exactly because a pair of distinct events can occur objectively "at the same time." But in Relativity, there is no absolute simultaneity, no moment diffused throughout all space. The totality of time is not the collection of all universal instants. Rather, the temporal structure is defined over individual events in spacetime. So we have to revisit questions about the dimensionality and topology of time.

Even without absolute simultaneity, time in Relativity can be characterized as locally one-dimensional. Consider, for example, event p in figure 27. In one sense, there are an infinite number of ways to move continuously in time from p: every time-like or light-like path through p represents a continuous set of events with a definite time order. But the light-cone structure at p partitions these paths into two distinct classes corresponding to the two light-cones. Any direction from p into or on one of the light-cones can be continuously shifted to any other direction into or on that light cone: picture smoothly twisting any arrow pointing "up" from p into any other such arrow. But one cannot smoothly transform an up-pointing arrow into a down-pointing arrow without going through some space-like directions. So just as one can go only to the east or to the west on the equator, the set of time-like and light-like directions from p divides into exactly two classes, which we call the future-directed and the past-directed.

The shift from absolute time, which is composed of instants, to the relativistic light-cone structure allows for some strange possibilities. Or at least, one can *mathematically* construct some very curious objects that look locally just like relativistic space-time. Some of these mathematical items seem clearly not to represent real physical possibilities for space-time, while others are a matter of fierce dispute. Let's begin with one so extreme that no one considers it to be a physical possibility.

Every event in a relativistic space-time has a light-cone that divides the time-like and light-like directions into two classes. Usually, these are called the future and the past light-cones. And in any mathematical model taken seriously by physicists, this distinction is *global*: having picked out the future light-cone at any one event, there is a unique way to determine the future light-cone at any other event. Referring back to figure 27, the upper lobe of the light-cone at p is the future light cone. But what about event t? Does settling which is the future light-cone at p settle the issue at t as well?

It should be obvious that it does. Imagine, for example, following the world-line of the Exterior Observer continuously from p to t. If we pick any future-directed arrow at p and carry it along that world-line, never allowing it to become space-like directed along the way, then it will point into the upper lobe of the light-cone when it arrives at t. And the same holds if one takes the path from p to r and then from r to t, as shown in the diagram, never allowing the direction of the arrow to become space-like. For any path from p to t (even a space-like path), the result will be the same: any direction pointing in the upper lobe at p will be continuously carried into a direction pointing into the upper lobe at t,   155

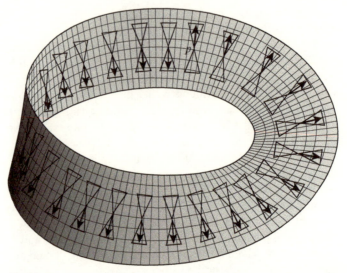

Fig. 29

as long as the direction is not allowed to become space-like along the way. So denominating one lobe at *p* the *future* light-cone and the other the *past* light-cone settles also the distinction between the future and past directions at all the other points of space-time. Such a space-time is called *temporally orientable*.

As a purely mathematical matter, one can describe non-orientable space-times. The most familiar non-orientable surface is the Möbius strip, and such a strip can be used to model a non-orientable two-dimensional space-time. Just draw light-cones so that a circuit around the strip results in the light-cone "flipping over": when one returns to the point of departure *p*, the direction originally called "future" now points into the opposite lobe (figure 29).

Locally, in any small enough patch, the structure depicted in figure 29 is isomorphic to a standard two-dimensional space-time (and the same trick can be played with a four-dimensional space-time). The structure is odd only with respect to its global topology. But the oddness is still fatal: no physicist would consider

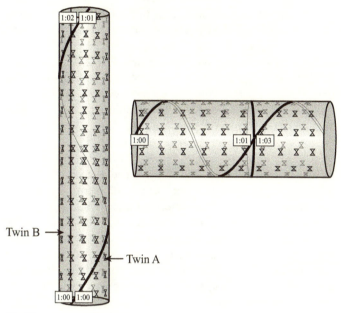

Fig. 30

figure 29 to be a representation of a possible way that space-time can be. The division into past and future light-cones ought to be consistent and global. Figure 29 is a mathematical curiosity with no physical import.

This is a critical lesson: even if we can describe a mathematical structure that everywhere looks *locally* like a possible space-time structure, it does not follow that the whole object corresponds to a physical possibility. There are less radical ways to manipulate the global topology of a mathematical representation, and it is an open question which of these depict ways that space-time might be. Figure 30 shows two other ways to "wrap up" a two-dimensional Minkowski space-time: one closes it in the space-like direction and the other in the time-like direction.

The space-time on the left in figure 30 has been rolled up—the official term is "compactified"—in a space-like direction. The circular cross-sections of the cylinder form "closed space-like

157

curves" that correspond, abstractly, to the equator on the earth. Just as one can circumnavigate the earth by heading constantly east or constantly west, one can circumnavigate this universe while occupying a straight, inertial time-like trajectory. Recall that both of these space-times look locally like Minkowski space-time: there is no space-time curvature anywhere. (The fact that a regular cylinder has no intrinsic curvature is easy to see since one can roll a flat piece of paper into a cylinder without stretching or deforming it.) So the space-like compactified Minkowski space-time allows us again to create a Twins Paradox situation without any acceleration. Both Twin A and Twin B in figure 30 follow unaccelerated paths, but Twin A will record less elapsed time between their successive meetings than Twin B.

In an obvious sense, the space-like compactified space-time is "spatially finite": there is only so much real estate for observers like Twin A and Twin B to occupy. Try as they might, they will never be farther away from one another than half the circumference of the cylinder, and as more observers are added, the situation gets more crowded.

The space-time on the left in figure 30 has some interesting properties, but the space-time on the right is profoundly alien. Here, the Minkowski space-time has been compactified in a time-like direction and the circular cross-sections of the cylinder form *closed time-like curves* (CTCs). This space-time, like the space-like compactified space-time, is temporally orientable: having settled which lobe of a single light-cone is the past light-cone and which the future, there is a unique partitioning of all light-cones into past and future. But in the time-like compactified space-time one can arrive back at the very same event having followed a continuous trajectory forward in time. Figure 30 shows the world-line of a single clock. When it shows 1:00, it is drifting inertially to the right in the diagram, and it continues to drift until it collides with an object at 1:01 and suddenly accelerates (the world-line bends). But the object it collides with is *the very same clock*. If you were an observer with the clock, the sequence of events would unfold like this: at 1:00 you are in inertial motion (seemingly "at rest") when you notice an object approaching. At 1:01, a collision occurs, sending the object (which happens to be a clock with an observer) on ahead. Looking backward (i.e., in the direction opposite from where the object appeared), you then see another clock with observer approaching "from behind." At 1:03, a second collision occurs, sending you on "forward" and stranding the colliding clock behind.

Of course, there is really only one collision in the history of this universe: the "earlier" stage of the clock colliding with its "later" stage. And there is, in this universe, only one material object: a single clock. But the temporal structure of the universe is closed in such a way that no event lies uniquely to the "past" or the "future" of any other event: every event lies in both the past and future light-cone of every other event, including itself. Unlike the space-like compactified space-time, there is no longer any objective sense in which any event is earlier rather than later than another. (It is amusing to trace out light rays in this universe: you can easily verify that our observer will always be able to see both past and future stages of himself, where "past" and "future" here refer to the time shown on the clock.)

In short, altering the topology of space-time in this way allows for a sort of time travel: an observer who always locally "goes forward" in time can return to an event he has already experienced. In a way, this is similar to how a traveler always going east on the equator can eventually come back home, but more deeply it is quite different: the traveler on the equator comes "back" to a different set of events than he left, and home may have changed in the meantime. The time traveler recurs to the very same event. There is no possibility, of course, of "changing the past": the collision in figure 30 only happens once, in one particular way, even though the observer experiences the event twice, playing a different role each time. This circumstance leads to some paradoxes, which we take up in the next section.

One puzzle about space-times with this peculiar temporal topology is illustrated by figure 30. Note that the clock at 1:00 could just as well—given the structure of the space-time and given that it is the only material object that exists—have continued forever without any collision, following an uninterrupted helical trajectory. Or, there could have been more collisions than are shown in figure 30. So there is a curious sort of underdetermination about what will happen to the clock, even though the physics may be completely deterministic in each small region of space-time.

Our initial concept of the temporal structure of the universe, enshrined in Newton's absolute time, is that the whole universe is in some state at any given time, and its state at later times is determined uniquely by its state at earlier times (if the laws of nature are deterministic). Even abandoning the notion of absolute simultaneity in Relativity does not undermine part of this fundamental picture. The relativistic replacement for a moment of absolute time is a *Cauchy surface*. A Cauchy surface is a portion of a relativistic space-time that every inextendible time-like curve intersects exactly once. So, for example, any circular or oblique cross-section (cut at less than 45°) of the space-like compactified space-time in figure 30 is a Cauchy surface: any time-like line, such as the world-line of Twin A or Twin B, will hit the cross-section exactly once if extended far enough. A relativistic space-time that admits of a Cauchy surface is called *globally hyperbolic*. And under the usual assumptions about relativistic space-times, any globally hyperbolic space-time not only admits of a single Cauchy surface: it admits of a foliation into a set of Cauchy surfaces. For example, our space-like compactified space-time can be sliced up into a stack of circular sections, or obliquely cut sections. If the laws of nature are deterministic and relativistically local, then any of these ways of "slicing" a space-time can play the same role as does the slicing of a classical space-time into simultaneity slices: the total physical state on any one of these slices, together with the laws of nature, determines the total physical state of the whole space-time.[2] Each of these different possible slicings of the space-time into Cauchy surfaces provides a way of thinking of the total state of the universe evolving forward in time.

But the time-like compactified space-time of figure 30 admits of no Cauchy surfaces. Since there are CTCs, a single time-like trajectory can intersect any region more than once, just as the trajectory of our clock intersects the region of the collision more

[2] For a thorough discussion, see Earman (1986).

than once. So there is no way to think of this universe as a whole as somehow evolving forward in time from a given state. This is a much more radically revisionary idea than the mere abandonment of absolute simultaneity.

Our example of the non-orientable Möbius space-time already served as a warning: not every well-defined mathematical object need represent a physical possibility. The "time travel" space-time of figure 30 is mathematically well-defined, and given our general principles we can describe how clocks and light rays would behave in such a situation. But none of that proves that real, physical space-time admits of such a possibility. Those who think that time essentially involves an asymmetric ordering of events, that no event can be both earlier and later than itself (in the sense that a time-like curve recurs to the same event) are free to reject the physical possibility of a space-time with CTCs. Unlike black holes, which will certainly form (according to the General Theory) when a sufficient amount of matter is confined to a small enough volume, there is no known physical condition that leads to the formation of CTCs.[3] The compactification of the models in figure 30 was done by hand: they have the topology they do because we decreed it, not because physics forced it upon them. Similarly, physicists can construct models that contain CTCs, but they do not naturally arise from known physical conditions.[4] So while it is fun to think about the problems of time travel, one should take the discussion a grain of salt: there is no solid

---

[3] See Earman, Smeenk, and Wüthrich (2009).

[4] A technical observation is in order here. In a certain sense, it is obvious that *any* space-time geometry on a manifold can be made consistent with Einstein's Field Equation. We can take an arbitrary metric, calculate the Einstein curvature tensor, then use the Field Equation to *define* a stress-energy tensor. So one way we judge that some geometries are "unphysical" is because we take some stress-energy tensors to be unphysical. One standard condition, called the Weak Energy Condition, essentially demands that in local Lorentz frames, the local energy density is nonnegative. Other such energy conditions can be formulated (cf. Hawking and Ellis [1973], sec. 4.3). So trying to find a physically plausible scenario for the formation of a CTC means finding one that satisfies reasonable energy conditions, yields a stable solution, and does not involve putting CTCs already in the initial conditions.

physical or empirical reason to think time travel into one's local past is physically possible.[5]

If the mathematical consistency of a model does not prove that it represents a real physical possibility, perhaps the question of the possibility of time travel can be settled in the opposite way by logic. It has long been supposed that time travel in the sense of CTCs would engender logical paradoxes or contradictions. But since a contradiction cannot be true, any supposition that leads to a contradiction cannot be true. Is the supposition of the possibility of time travel like that?

The standard time-travel paradox involves travelling to the past and doing something that ensures that one cannot possibly have ever travelled to the past—an obvious contradiction. One dramatic example involves killing one's own grandfather before he has had any children. If your grandfather had not had any children then obviously you would not have existed: one of your parents would not have existed. Or, in the same vein, one could travel to the past and kill oneself as a child, assuring that one would never grow up to travel to the past. Clearly, these sorts of stories are inconsistent, and hence impossible. If time travel somehow implies these sorts of scenarios, then time travel is also impossible.

But once we get down to the nitty-gritty physical details, it turns out to be very difficult to describe how someone could undertake a time-travel mission that only yields contradictory or paradoxical outcomes. In the auto-suicide case, for example, one reasons: if I go back and kill my younger self, then I won't be here to go back, which is a contradiction. But if I don't go back and kill myself, there is nothing to prevent my doing it, so I will: another

[5] The reader might wonder: if Minkowski space-time can be compactified in a time-like direction and in a space-like direction, what about compactifying it in a light-like direction, forming closed light-like curves? This is also mathematically possible and is easy to do in the two-dimensional case. I leave it as an exercise. Note that neither of the models in figure 30 has any closed light-like geodesics.

contradiction. Now at a pedestrian level, the second claim is obviously highly questionable: just because you do not go back and kill your younger self, it does not at all follow that nothing could prevent you from doing so if you try. All sorts of things could prevent the success of the auto-suicide attempt: a change of heart, a heroic intervention by your parents, and so on. These are all *logically* possible resolutions, as is a resolution where you are prevented from time-travelling in the first place by a change of heart or the heroic intervention of your (now frail) parents. These sorts of *deus ex machina* scenarios, though, smack of the arbitrary: logical consistency is saved, but not by any guiding principle.

A much more satisfying resolution arises from the observation that under certain very plausible physical constraints, it is provable that there will always be a consistent way to play out the scenario without any outside intervention: the self-interacting system provides its own consistent possibilities. For example: you go back in time intent on assassinating your younger self. You make sure to go back to a day in your youth when your parents are away; you pump yourself up with rage to ensure there are no second thoughts, and so on. You get back and see your infant self in the cradle, unprotected and helpless. You raise your gun to fire . . . but there is a tremble in your hand, and instead of killing your younger self, you merely inflict a shoulder wound. In fact, the nerve damage caused by that shot affects you all your life, leading in particular to the exact tremble on the fateful day. Paradox and inconsistency are avoided, but without invoking any intervention from outside the self-interacting system. The physics of that system alone allows for a consistent story.

This closed causal loop,

tremble → missed shot → nerve damage → tremble,

may seem to be as miraculous as parental intervention, but as a purely mathematical matter one can prove under some general assumptions that such a consistent evolution of a closed, self-interacting system always exists.[6] In general, we need only to assume that the physics is deterministic and continuous, and that

---

[6] See Clarke (1977); Arntzenius and Maudlin (2009).

the space of possible physical states of a system has a certain simple geometrical form. Physicists have studied this same problem, for billiard balls being sent back in time through wormholes to interact with their younger selves, and found exactly the same result.[7] In fact, what one finds under plausible assumptions is that far from there not being *any* solutions to the physical equations in these cases, there are typically *many* solutions. The situation is exactly like our clock in figure 30: as far as the laws of physics are concerned, the clock could collide with itself when it shows 1:01, as is depicted, or it could collide with itself in the same way earlier or later, many times or not at all.

It should be emphasized again that time travel scenarios never provide the possibility to go back in time and *change* the past, that is, to make the past different from what it actually was. This is just logically impossible. Since you only had one childhood, it either contained a near-fatal assault that left you wounded or it did not. If it did, and if time travel is possible, then perhaps the assailant was an older version of yourself. If it didn't, then it didn't, even if, due to time travel, your later self was present in your childhood. The past—like the present and future— only happens once, and happens in a certain way. This is not a deep metaphysical principle. To say an event happened one way "the first time around" and another way "the second time around" makes no sense, because there is no first and second time.

This triviality is hard to see because many fictional time-travel stories are inconsistent and hence impossible. Marty McFly, in *Back to the Future*, cannot *change* the way that his parents actually met, for the simple reason that they met as they did. If he can go back in time, he can play a role in their meeting, he can *influence* how it happens. But it can't be that first there was one past set of events without him and then a second set of events with him, since there was only one set of events.

Inconsistent time travel stories are usually really concerned with the radical *contingency* of life: the fact that how one's life plays out depends on many small factors beyond one's knowledge and control. This radical contingency is illustrated in Robert Frost's poem *The Road Not Taken*, and in movies that show how a life

[7] Echeverria, Klinkhammer, and Thorne (1991).

would have gone had some trivial-seeming event happened differently. Examples of this genre are *Sliding Doors* and *Run, Lola, Run* and Krystof Kieslowski's *Blind Chance*. These movies tell the same story multiple times, with a small variation leading to a different ending. Time travel stories are often really concerned with this sort of counterfactual question: what would have happened if . . . ? It is an interesting and important question, but trying to address it by time travel leads to incoherence.

There are also some consistent time-travel tales. *Twelve Monkeys* (based on Chris Marker's *La Jetée*) is a prime example, as is the *Time's Arrow* episode from *Star Trek: The Next Generation*. Perhaps the most extreme example is the story "All You Zombies" by Robert Heinlein, in which the main character turns out to be both of his own parents, all in a logically impeccable way. These stories can give no insight into what might have been, since they simply recount the particular details of what actually happened. One might think that these consistent time-travel stories could not be filled out in all detail, but the mathematical results mentioned above show this isn't true.

The mere logical consistency of time travel does not show that it is in any serious sense either physically or metaphysically possible. "Richard Nixon was a ham sandwich" is a *logically* consistent proposition. It is certainly false. But more importantly, one cannot describe a *possible* scenario that would make it true. It just *couldn't* have been true no matter what: the world could not have been arranged in such a way as to both contain Nixon and to have that very man be a ham sandwich. "Logical possibility" is not really a species of possibility: it merely means that logic alone does not forbid the truth of a proposition. Similarly, logic and mathematics do not rule out time travel, as our model in figure 30 shows. But just as the nature of Nixon precludes him from being a ham sandwich, perhaps the nature of time precludes it from going around in a circle. It is difficult to know how to go about addressing a question like this.

THE DIRECTION OF TIME

The self-generating protagonist of "All You Zombies" makes us queasy for the same reason that the closed causal loop from the tremble of the shooter's hand back to itself does. Intuitively, causation is an asymmetric and hence irreflexive relation: if A causes B then B cannot cause A, so nothing can produce itself. And intuitively, the asymmetry of causation is itself parasitic on a fundamental asymmetry of time. The future arises from the past. The past-to-future direction of time is fundamentally unlike the future-to-past direction in a way that has no spatial analog. There are only two directions along the equator, but the equator itself no more runs from east to west than it does from west to east. Time, in contrast, passes from past to future, and we are all inevitably headed toward the grave and away from the nursery.

The direction of time is embedded so deeply into our language and concepts that it is impossible to expunge. Most verbs are time-directed: the difference between a rock falling and the rock rising is determined by which direction of time is toward the future and which toward the past. The words "to" and "from" have their usual application by reference to time: processes run from earlier states to later ones. We remember the past and anticipate the future. Our present actions can influence the course of the future but are impotent to influence the past.

To say that the past-to-future direction of time is intrinsically different from the future-to-past direction is not the same as saying that the future must be qualitatively different from the past. The physicist Fred Hoyle once proposed a "steady state" theory of the universe: the universe as a whole is and always has been expanding, with galaxies moving ever farther apart from one another. But in Hoyle's theory, the average density of matter in the universe does not go down: new matter is created in the empty spaces at just the right rate to keep the density constant. In this model, all times—future, present, and past—are qualitatively the same: hence "steady state." But there is still a fundamental direction of time, in virtue of which it is correct to say that the universe is *expanding* and new matter being *created*, rather than *contracting* with matter being *destroyed*.

Our world is filled with processes that have an evident time direction. Ice cubes in hot water, isolated from outside influence, melt and produce lukewarm water. Lukewarm water, isolated from outside influence, never spontaneously segregates into ice cubes and hot water. But the laws of physics seem to allow the latter process just as much as the former. Once again, our very description of the two processes presupposes a time direction: the only difference between melting and freezing is the direction of time. Similarly, our description of a black hole in figure 27 presupposes a direction of time: it is because the direction to the future is represented by up on the diagram rather than down that we say things *fall into* and *never escape* the event horizon rather than *are ejected from* and *cannot remain within* it. Just as the fundamental laws seem to allow both the melting and spontaneous freezing of ice, so General Relativity allows for both black holes and their time reverses. And just as ice often melts but never spontaneously forms in lukewarm water, so there seem to be many black holes in the universe but no time-reverses of them.

There are, then, two distinct questions to ask about the direction of time. One is why there seem to be processes that never occur even though their time-reverses regularly occur. When asked in terms of, say, spontaneous melting and spontaneous freezing, this question presupposes that there is a fundamental difference between the past-to-future and future-to-past direction. But some philosophers and physicists have drawn a much more startling conclusion from the fact that the laws of physics seem to allow both sorts of processes: they have concluded that there is no fundamental distinction between the two time directions at all. For example, Paul Horwich asserts, "I'll argue that the current empirical evidence indicates that time itself is intrinsically symmetric."[8] Huw Price states: "[I]t has not been properly appreciated that we have no right to assume that entropy *increases* rather than *decreases*, for example. What is objective is that there is an entropy gradient over time, not that the universe 'moves' on this gradient in one direction rather than another."[9]

Consider the implications of these claims. The difference between *increasing* and *decreasing* is just the direction of time: entropy is increasing at a moment if it gets greater toward the future and decreasing if its gets greater toward the past. So if there is

[8] Horwich (1987), p. 38.
[9] Price (1996), p. 48.

no objective fact about whether entropy is increasing or decreasing anywhere, then there is equally no objective fact about any proposition that uses a time-directed verb. We are not, in any deep objective sense, headed "toward" our death and "away from" our birth, we are not, in any objective sense, getting older rather than younger. Indeed, the whole idea that time "passes" at all is, on this view, some sort of illusion. Later states of the universe do not, fundamentally, arise from earlier ones any more than earlier ones arise from later ones. The whole notion of cause and effect, or production of one thing from another, is equally illusory.

Here is one way to see what is at stake in the debate. If the future-to-past temporal direction is really qualitatively identical to the past-to-future direction, then there is nothing obviously physically impossible about the non-orientable Möbius space-time depicted in figure 29. Apart from the distinction between the two time directions, the light-cone structure of the space-time is perfectly smooth everywhere. But if the two temporal directions are fundamentally distinct, then such a non-orientable structure cannot represent a possible space-time, as it is inconsistent with the nature of time. Since no physicist or philosopher has ever suggested the physical possibility of a temporally non-orientable space-time, the notion that the two directions are fundamentally different seems to have been at least tacitly assumed.

It is not easy in the literature to disentangle these two questions from one another. The question of the de facto directionality of processes—why ice cubes spontaneously melt in hot water but lukewarm water does not spontaneously segregate itself into ice cubes and hot water—is addressed by statistical physics, which is a topic of volume 2. The idea that time has no intrinsic directionality at all is harder to evaluate. One could be lulled into this idea by an unreflective use of space-time *diagrams*, which are purely spatial objects with no intrinsic directionality. The past-to-future direction represented by such a diagram must be indicated by some extra convention, since the medium of the representation itself has no such asymmetry. But equally, the difference between the representation of space-like and time-like directions on the diagram must be indicated by a convention. Similarly, purely *mathematical* representations of space-time require many

conventions, such as whether time-like Intervals are to be represented by real or imaginary numbers, or whether a time coordinate increases or decreases in the past-to-future direction. None
of this even vaguely suggests that the two time directions are not
fundamentally different, any more than it suggests that time-like
directions do not differ intrinsically from space-like directions.

I have addressed some of the arguments about the passage of
time in "On the Passage of Time."[10] But since these issues cannot
be properly discussed without some understanding of matter and
the laws that govern matter and statistical physics, a more com-
prehensive discussion of the direction of time will have to await
volume 2.

---

[10] Maudlin (2007), chap. 4.

Appendix: Some Problems in Special Relativistic Physics

I HAVE PRESENTED Special Relativity in a somewhat atypical way, focusing directly on the geometry of Minkowski space-time and only secondarily on Lorentz coordinates. The coordinates come in merely as a convenience, in terms of which the Interval between events can be easily calculated. This approach makes certain sorts of problems rather simple to solve without ever having to write down the Lorentz transformations. I have also relied heavily on the use of space-time diagrams, pointing out the similarities and differences between Euclidean space and Minkowski space-time.

The key to solving these problems is to use a convenient Lorentz frame. By a convenient frame, I mean one in which the calculations will be easy to do: we can use any frame at all and get the right results, but in some frames the math is simpler. Choosing a convenient frame usually means deciding to treat some inertially moving massive object as "at rest." The world-line of this object will be a vertical line in your space-time diagram. If this object is a clock, it will tick off the t-coordinate values in this Lorentz frame.

Once you have chosen which object to treat as "at rest," draw a space-time diagram. Remember: inertially moving objects and light rays in a vacuum are represented by straight lines, and the trajectories of light rays are *always* set at a 45° angle in the diagram. Once you have the diagram, put in any information you have about the Intervals between events. This information is the same in all Lorentz frames. So, for example, for any light ray, show that $I = 0$. And if a clock is in the problem and its readings are given, recall that a clock measures I along its trajectory.

With all of this information, you can determine the coordinates of events in the chosen frame. You have the general "pseudo-Pythagorean" formula for calculating I in any reference frame:

$$I(p,q) = \sqrt{\left(T(p) - T(q)\right)^2 - \left(X(p) - X(q)\right)^2 - \left(Y(p) - Y(q)\right)^2 - \left(Z(p) - Z(q)\right)^2},$$

which is the source of your equations. These are two-dimensional problems, so you only ever need a t-coordinate and an x-coordinate for each event. The difference in the t-coordinates in a frame give you "time elapsed between events" in that frame, and the difference in x-coordinates give "distance between events" in that frame. From the "time elapsed" and "distance" in a frame, you can calculate the "speed" in that frame. But always keep in mind that "time elapsed," "distance" and "speed" are *frame-dependent* quantities: they vary from one frame to another. Calculating them in one frame does not tell you what they are in another frame. The Interval, in contrast, is a frame-independent quantity: it is the same in all Lorentz frames.

You can apply the "pseudo-Pythagorean formula" $I = \sqrt{\Delta t^2 - \Delta x^2}$ to "right triangles" in the diagram to calculate the Interval length of the hypotenuse, but only on triangles where one leg (Δt) is vertical and the other (Δx) is horizontal. The key to solving many problems is drawing in additional lines to form the appropriate right triangles. Each triangle provides an equation.

The path of any inertially moving object will be given by a linear equation in a Lorentz frame. In a two-dimensional problem, using only t and x, the equation will be of the form $x = vt + C$, where v is the velocity of the object in this frame and C is some constant.

1. Clocks A and B start out synchronized next to each other. At exactly 12:00 (on each clock) they separate, moving inertially in opposite directions. When it shows 12:01, clock A sends out a light ray. When it receives the light ray, clock B shows 12:02. Using only the definition of the interval and facts about clocks and light rays, determine the following.

   a) In the reference frame of clock A, when does the light ray arrive at clock B?
   b) In the reference frame of clock A, how far apart are the clocks when the light ray arrives at clock B?
   c) In the reference frame of clock A, how fast is clock B moving?
   d) In the reference frame of clock A, how far apart were the clocks when clock A emitted the light ray?

e) In the reference frame of clock B, when does clock A emit the ray?

f) In the reference frame of clock B, how far apart are the clocks when the light ray arrives at clock B?

g) In the reference frame of clock B, how far apart were the clocks when A emitted the ray?

h) From the answers given above, calculate the amount by which each clock judges the other to be slowed down.

i) The usual time-dilation formula is says that a clock moving at speed v is slowed down by a factor of $\frac{1}{\sqrt{1-v^2/c^2}}$, where c is the speed of light. Compare this with the result of part h. Recall that in a Lorentz frame, as we have defined it, $c = 1$.

(Some of these problems can be solved more easily if you remember that the velocity of clock A relative to clock B equals the negative of the velocity of clock B relative to clock A.)

2. Two twins, Sam and Sue, start out together on the earth. Sue blasts off in a spaceship and then travels inertially until her clocks show that a year has passed. She then reverses her direction and travels inertially again. When she arrives back on the earth, two more years have passed (on her clocks) for a total of three. When she gets back, Sam is four years older than he was when she left.

a) According to Sam's reference frame, when does Sue turn around?

b) According to Sam's reference frame, how far apart are the twins when Sue turns around?

c) According to Sam's reference frame, how fast does Sue go out? How fast does she come back?

d) According to Sue's inertial frame as she goes out, how far apart are the twins when she turns around?

e) According to Sue's inertial frame as she comes back, how far apart are the twins when she turns around?

3. In order to prove that special relativity does not require coordinates, consider how to solve a problem using

Appendix

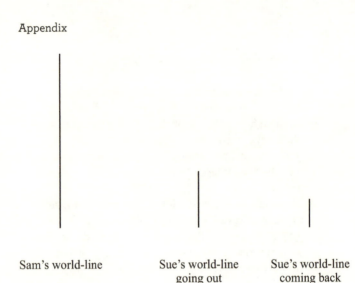

Sam's world-line   Sue's world-line   Sue's world-line
                          going out   coming back

Fig. A.1

only a Minkowski straightedge and Minkowski compass. That is, suppose that you can draw straight lines
in space-time using the straightedge and that the compass measures out loci of equal Invariant Interval from
a given point. If you center your compass at p and set it
to some Interval, it draws the set of points that are that
Interval from p. Describe in detail how you would use
your two instruments to solve the following problem.

As in problem 2, Sue travels out and back while Sam
stays on the earth. The lengths of their world-lines are as
shown in figure A.1. At the moment she turns around,
Sue emits a light ray. Using your instruments, explain
how to construct the length of the world-line that represents the time on the earth between the reception of
Sue's signal and Sue's return.

Solution to Parts a–d of Problem 1   Begin by drawing a space-
time diagram. Since parts a–d ask for quantities "in the reference
frame of clock A," clock A will be represented by a vertical line on
174   the diagram. The light ray is a line at 45°, and clock B is an angled
straight line, since it too is in inertial motion. The readings on

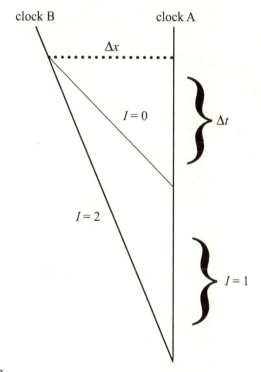

Fig. A.2

the clocks supply the relevant Intervals, and the Interval of the light ray is 0. Once you have the objects drawn in, look for right triangles. If necessary, draw extra lines to form such triangles.

The resulting diagram is given in figure A.2, with the extra line indicated as Δx. We can see two right triangles in the diagram. One has sides Δx and Δt with the light ray as hypotenuse, and the other has sides Δx and $1 + \Delta t$, with clock B as hypotenuse. The vertical side of the second triangle can be written as $1 + \Delta t$ because clock A measures the t coordinate in this frame, so $I = 1$ for this clock translates directly into a measure of the change of t-coordinate along this path. Clock B, in contrast, does not indicate the t-coordinate in this frame: it will appear "time dilated." Of course, clock B measures the t-coordinate in its own rest frame, which is relevant to the later parts of the problem.

Our two right triangles yield two equations:

$$0 = \sqrt{\Delta t^2 - \Delta x^2}$$
$$2 = \sqrt{(1 + \Delta t)^2 - \Delta x^2}.$$

Squaring both gives

$$0 = \Delta t^2 - \Delta x^2$$
$$4 = (1 + \Delta t)^2 - \Delta x^2$$

And expanding the square in the second equation gives

$$4 = 1 + 2\Delta t + \Delta t^2 - \Delta x^2.$$

Since by the first equation, $\Delta t^2 - \Delta x^2 = 0$, we derive

$$4 = 1 + 2\Delta t$$

$$\Delta t = 3/2.$$

So the t-coordinate, in this frame, of the point where the light ray reaches clock B is $1 + 3/2 = 5/2$. According to this frame, the light ray reaches clock B 2½ minutes after the clocks separate.

Already we can see the "time-dilation effect" in this frame: according to the frame, 2½ minutes have elapsed but clock B only shows 2 minutes.

The distance between the clocks, in this frame, when the light ray reaches clock B is just Δx on the diagram. But by our first equation, $\Delta x = \Delta t$. (This holds for any light ray.) So the distance is $3/2$: the two clocks are 1½ light-minutes apart when the light ray reaches clock B, according to clock A's frame.

Since, according to this frame, clock B has gone 1½ light-minutes in 2½ minutes, its speed is $\frac{3/2}{5/2} = \frac{3}{5}$ the speed of light. This indicates how fast a clock must go in a frame to get the level of frame-dependent time dilation we have calculated.

Once we have the speed of clock B in this frame, we can calculate how far apart the clocks were when the light ray was emitted. Since (in this frame) clock B moves at 3/5 the speed of light and (in this frame) the light ray is emitted 1 minute after the clocks separate, in this frame the clocks are judged to be 3/5 light-minute apart at that moment. As you will see, all of these quantities are frame-dependent: a very different story is told in the rest frame of clock B.

References

Alexander, H. G., ed. 1956. *The Leibniz-Clarke Correspondence*. New York, NY: Barnes and Noble.

Arntzenius, Frank, and Tim Maudlin. 2009. "Time Travel and Modern Physics." In E. N. Zalta, ed., *Stanford Encyclopedia of Philosophy*. http://plato.stanford.edu/entries/time-travel-phys/.

Baez, John, and Emory Bunn. 2006. "The Meaning of Einstein's Equation." http://math.ucr.edu/home/baez/einstein/.

Barbour, Julian. 2000. *The End of Time*. New York, NY: Oxford University Press.

———. 2001. *The Discovery of Dynamics*. New York, NY: Oxford University Press.

Barbour, Julian, and Bruno Bertotti. 1982. "Mach's Principle and the Structure of Dynamical Theories." *Proceedings of the Royal Society A* 382 (1783): 295–306.

Bell, J. S. 2008. *Speakable and Unspeakable in Quantum Mechanics*. 2d ed. Cambridge, UK: Cambridge University Press.

Brown, Harvey. 2006. *Physical Relativity*. Oxford, UK: Oxford University Press.

Butterfield, Jeremy. 1989. "The Hole Story." *British Journal for the Philosophy of Science* **40**:1–28.

Carroll, Sean. 2004. *Spacetime and Geometry: An Introduction to General Relativity*. Boston, MA: Addison-Wesley.

Clarke, C. J. S. 1977. "Time in General Relativity." In J. Earman, C. Glymour, and J. Stachel, eds., *Foundations of Space-time Theories*. Minnesota Studies in the Philosophy of Science, vol. 8. Minneapolis, MN: University of Minnesota Press.

DiSalle, Robert. 1991. "Conventionalism and the Origins of the Inertial Frame Concept." In A. Fine, M. Forbes, and L. Wessels, eds., *PSA 1990*, vol. 2. East Lansing MI: Philosophy of Science Association.

Earman, John. 1986. *A Primer on Determinism*. Dordrecht, NL: Reidel.

———. 1989. *World Enough and Space-Time*. Cambridge, MA: MIT Press.

———. 1995. *Bangs, Crunches, Whimpers, and Shrieks: Singularities and Acausalities in Relativistic Space-Times*. Oxford, UK: Oxford University Press.

References

Earman, John, and John Norton. 1987. "What Price Space-Time Substantivalism? The Hole Story." *British Journal for the Philosophy of Science* **38**:515–525.

Earman, John, Christopher Smeenk, and Christian Wüthrich. 2009. "Do the Laws of Physics Forbid the Operation of a Time Machine?" *Synthese* **169** (1): 91–124.

Echeverria, Fernando, Gunnar Klinkhammer, and Kip S. Thorne. 1991. "Billiard Ball in Wormhole Spacetimes with Closed Timelike Curves: Classical Theory." *Physical Review D* **44** (4): 1077–1099.

Einstein, Albert. 1982. "Autobiographical Notes." In P. A. Schilpp, ed., *Albert Einstein: Philosopher-Scientist*, vol. 1. LaSalle, IL: Open Court.

Feynman, Richard, Robert Leighton, and Mathew Sands. 1975. *The Feynman Lectures on Physics*. Reading, MA: Addison-Wesley.

Friedman, Michael. 1986. *Foundations of Space-Time Theories: Relativistic Physics and Philosophy of Science*. Princeton, NJ: Princeton University Press.

Galileo. 1967. *Dialogue concerning the Two Chief World Systems*, trans. by Stillman Drake. Berkeley, CA: University of California Press.

———. 1974. *Two New Sciences*, trans. by Stillman Drake. Madison, WI: University of Wisconsin Press.

Geroch, Robert. 1978. *General Relativity from A to B*. Chicago, IL: University of Chicago Press.

Goldstein, Herbert. 1981. *Classical Mechanics*, 2d ed. Reading, MA: Addison-Wesley.

Hawking, S. W., and G. F. R. Ellis. 1973. *The Large-Scale Structure of Space-Time*. Cambridge, UK: Cambridge University Press.

Hoefer, Carl, and Nancy Cartwright. 1994. "Substantivalism and the Hole Argument." In John Earman, Al Janis, Gerald Massey, and Nicholas Recher, eds., *Philosophical Problems of the Internal and External Worlds*. Pittsburgh, PA: University of Pittsburgh Press.

Horwich, Paul. 1987. *Asymmetries in Time*. Cambridge, MA: MIT Press.

Huggett, Nick, ed. 1999. *Space from Zeno to Einstein*. Cambridge, MA: Bradford.

Kripke, Saul. 1980. *Naming and Necessity*. Cambridge, MA: Harvard University Press.

Lewis, David. 1986. *On the Plurality of Worlds*. Oxford: Basil Blackwell.

Maudlin, Tim. 1989. "The Essence of Space-Time." In Arthur Fine and Jarrett Leplin, eds., *Proceedings of the Philosophy of Science Associa-*

tion Meetings, 1988, vol. 2. East Lansing, MI: Philosophy of Science Association.

———. 2007. *The Metaphysics Within Physics.* Oxford, UK: Oxford University Press.

———. 2010. "The Geometry of Space-Time." In *Aristotelian Society Supplementary Volume 84.* London: Aristotelian Society.

Misner, Charles, Kip Thorne, and John Wheeler. 1983. *Gravitation.* San Francisco, CA: W. H. Freeman.

Newton, Isaac. 1934. *Principia*, trans. by Andrew Motte, rev. by Florian Cajoli. 2 vols. Berkeley, CA: University of California Press.

———. 1962. *Unpublished Papers of Isaac Newton: A Selection from the Portsmouth Collection in the University Library, Cambridge*, transl. and ed. by A. R. Hall and M. B. Hall. Cambridge, UK: Cambridge University Press.

Norton, John. 1992. "Einstein, Nordström and the Early Demise of Lorentz Covariant, Scalar Theories of Gravitation." *Archive for History of Exact Sciences* **45**: 17–94.

———. 2008. "The Hole Argument." In Edward N. Zalta, ed., *Stanford Encyclopedia of Philosophy.* http://plato.stanford.edu/archives/win2008/entries/spacetime-holearg/.

Price, Huw. 1996. *Time's Arrow and Archimedes' Point.* Oxford, UK: Oxford University Press.

Rindler, Wolfgang. 1977. *Essential Relativity.* New York, NY: Springer-Verlag.

Rodriguez-Pereyra, Gonzolo. 1999. "Leibniz's Argument for the Identity of Indiscernibles in His Correspondence with Clarke." *Australasian Journal of Philosophy* **77** (4): 429–438.

Rynasiewicz, Robert. 1995. "By Their Properties, Causes and Effects: Newton's Scholium on Time, Space, Place and Motion." *Studies in History and Philosophy of Science* **26**. "Part I: The Text," 133–153. "Part II: The Context," 295–321.

Sklar, Lawrence. 1977. *Space, Time, and Spacetime*, Berkeley, CA: University of California Press.

Wald, Robert. 1984. *General Relativity.* Chicago, IL: University of Chicago Press.

Index

Index

Printed in the USA
CPSIA information can be obtained
at www.ICGtesting.com
JSHW021514140823
46506JS00008B/83